An Introduction to
WASTEWATER TREATMENT

by Tom Rae, BA, FCIWEM (Dip), I. ENG, AIBA

The Chartered Institution of Water and Environmental Management

First published (by IWPC) 1960
Second edition 1968
Revised 1986

**The Chartered Institution of Water and
Environmental Management, 1998
ISBN 1 870752 34 1**

Published for CIWEM by Terence Dalton Publishers Ltd

FOREWORD

The Chartered Institution of Water and Environmental Management (CIWEM) is a multi-disciplinary body and examining organization for professionals engaged in environmental management. CIWEM was formed in 1987 by the unification of three eminent organizations, the Institution of Public Health Engineers, the Institution of Water Engineers and Scientists and the Institute of Water Pollution Control, each having a history of about 100 years. It is now the leading professional body in this sector, having been granted a Royal Charter in 1995. CIWEM has almost 12 000 members working on international, national and local projects in operational management, education and research.

Amongst the Aims are the promotion of learning, in order to advance standards, and the dissemination of information through publications, for the wider public benefit. However, we also aim to make the learning enjoyable because we should learn through enjoyment and obtain enjoyment through learning.

This publication is the eighth in a series of Introductory Booklets – each on an aspect of environmental management of topical interest. Although it is intended as an introduction to the subject, and has been written as a general guide to the interested lay person, the book provides a comprehensive summary of the situation in the UK, and should also be useful for those embarking on a career in the field. A further CIWEM series entitled *Handbooks of UK Wastewater Practice* provides much more detailed information on the various unit processes which are employed in sewage treatment.

The Institution wishes to record its thanks to those who have contributed to the production of this booklet, and in particular to *Tom Rae* who has been responsible for the preparation of the text and to *Ted Bruce* the Editor.

Please enjoy the publication and make good use of the learning which you obtain from it. Share it with your colleagues.

Peter Matthews
President

September 1998

ACKNOWLEDGEMENTS

The Institution gratefully acknowledges permission from the following to reproduce illustrations:

ACWa Services Ltd. (Fig. 3)
Anglian Water Ltd. (Plate 5)
Birse Construction Ltd. (Plates 1 & 6)
Biwater International Ltd. (Plate 2)
CIWEM Library (Plate 8)
H. Leverton Limited. (Plate 10)
Severn Trent Water Ltd. (Plate 7)
E. Tomkinson Co. Ltd. (Plate 4)
Tuke & Bell Ltd. (Plate 3)
USF Limited. (Plate 9)

PREFACE

This introductory booklet is intended for a wide readership, including staff in the water and environment industry who may not be fully involved in the subject area but need a basic appreciation of the processes and philosophy of sewage treatment, e.g. teachers, undergraduates and senior school students, and lay persons who have a personal interest in the subject.

The basic processes of sewage treatment have remained the same for many years, with frequent modifications and improvements by the many dedicated professionals in the industry. The impact of privatization, the separation of the regulatory function from the operators and the strong influence of EU and UK legislation, as well as a greater environmental awareness by the general public, have generated the funding and hence the technology to achieve consistent higher-quality effluents.

The diverse disciplines now involved in sewage treatment, coupled with the increased technology and skills required, have resulted in many interesting and rewarding opportunities in this sector of the water and environment industry.

I am pleased to record my grateful thanks to the many companies which have provided information to help prepare the new edition of this booklet and, in particular, to the CIWEM members who have willingly commented on early drafts of the booklet. Particular thanks are due to Alan Bruce, Tony Gilks, Malcolm Haigh, Malcolm Helm, Roger Holdsworth, Nick Sambidge, Colin South and Alan Tong.

Tom Rae

September 1998

CONTENTS

	Page
Foreword	iii
Acknowledgements	iv
Preface	v
List of Figures and Plates	ix

1. **Introduction** 1

2. **Historical Development** 4
   Nineteenth Century 4
   Royal Commission on Sewage Disposal 6
   Recent Legislation 6
   European Union Legislation 7
   Conservation and Re-use 8

3. **Reasons for Wastewater Treatment** 10
   Measurement of Water Quality 10
   Effects on Aquatic Life 13

4. **Domestic and Industrial Effluents** 14
   General Characteristics 14
   Treatment Processes 15

5. **Preliminary Treatment** 18
   Screening 18
   Grit Separation 20
   Storm-Sewage Treatment 20

6. **Primary Treatment** 22
   Sedimentation 22

7. **Secondary Treatment I** 25
   Biological Filtration 25
   Alternative Methods of Operation 27
   Other 'Captive-Film' Systems 29

8. Secondary Treatment II 31
   Activated-Sludge Process 31
   Alternative Methods of Operation 34
   Comparison with Biological Filtration 36

9. Tertiary Treatment 37
   General Considerations 37
   Tertiary Treatment Methods 38

10. Sludge Treatment and Disposal 43
   Sludge Characteristics 43
   Sludge Treatment 44
   Sludge Disposal 51

11. Utilization of Products of Wastewater Treatment 53
   Heat and Energy from Sludge Gas 54
   Soil Conditioning 55
   Heat and Energy from Sludge 55

12. Control, Training and Administration in UK 57
   Control of Sewage-Treatment Processes 57
   Administration in England and Wales 58
   Administration in Scotland 59
   Administration in Northern Ireland 59

References 61

Index 63

FIGURES

		Page
1.	Simplified representation of a sewerage system and sewage-treatment works	2
2.	Basic layout of a medium-sized sewage-treatment plant with biological filtration	16
3.	Example of fine screen (spiral sieve screen)	19
4.	Diagrammatic layout of a biological-filter plant with recirculation of final effluent	28
5.	Diagrammatic layout of an activated-sludge plant	31
6.	General arrangement of a fluidized-bed incineration plant	50

PLATES

1.	Aerial view of a typical medium-sized sewage-treatment works	11
2.	Radial-flow tank showing arrangement of sludge scrapers	23
3.	Circular biological filters	26
4.	Rotating biological contactor	29
5.	Diffused-air activated-sludge plant	32
6.	Vertically mounted surface aerator	33
7.	New reedbed at Norton Lindsey, near Stratford-on-Avon	40
8.	Sludge digestion plant showing glass-coated steel tanks with gas mixing and plastic membrane gas-holder	46
9.	Filter press	48
10.	Combined heat and power installation	54

1. INTRODUCTION

As part of our every-day lives, we all pollute water. Every time we wash, shower or bathe, brush our teeth, use the toilet, do the washing-up, operate the washing machine or clean the car, we contaminate the high-quality mains water which is supplied to our homes. The amount of water that we individually pollute in this way averages about 135 litres/day, and a typical household in the UK will produce about 475 litres of wastewater (or sewage) each day. In most cases, we conveniently dispose of this water by discharging it through the household drains and into the public sewers beneath the streets (Fig 1).

Once the polluted household water is in the sewer, it becomes technically known as 'sewage' or, to use the term now increasingly used in the context of European Union legislation, 'urban wastewater'. As it flows through the sewer (or sewerage)** system, it becomes mixed with the sewage from our houses and from offices, shops and other establishments. In most cases the sewer will also receive wastewater discharges from factories and other industrial premises. Industrial discharges may equal, or even exceed (in volume), the total amount of sewage which is produced from households in the same catchment area. Eventually, the combined sewage flow from the various sources will be conducted, by gravity, (or sometimes assisted by pumping) to a low-lying area in the catchment which is usually close to a natural watercourse or estuary or, in coastal areas, to the sea.

Without treatment, the crude sewage would inevitably flow into the natural waters and cause pollution. The severity of the pollution would depend upon various factors and, particularly in the case of inland rivers, pollution by untreated sewage would cause the death of fish and other aquatic life, and would cause the water to become septic, malodorous and a potential source of disease.

** It is important to distinguish between the term 'sewage' which is used to describe the liquid carried in a sewer, and 'sewerage' (or sewerage system) which is used to describe the network of pipes serving as sewers.
In this book, the term 'sewage' is used throughout rather than 'urban wastewater', although in most cases, the two terms have identical meaning.

Fortunately, in the UK, the pollution (by sewage) of our natural watercourses, estuaries and coastal waters, is largely avoided by the widespread provision of sewage-treatment works. Such works are designed and operated to treat crude sewage to a degree which is adequate to ensure that the resultant effluent (which is discharged from the works) will not cause unacceptable pollution of the receiving waters.

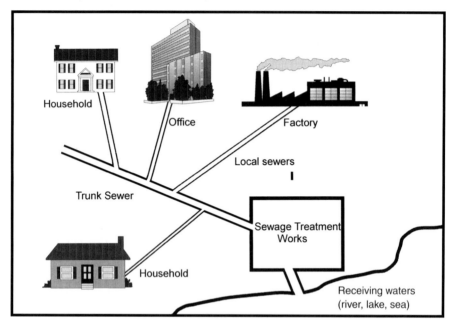

Fig. 1. Simplified representation of sewerage system and sewage-treatment works

The provision of a comprehensive sewerage system and of an adequate number of sewage-treatment works is a basic requirement for sustainable development. In the UK, 97% of the population is connected to a sewer; there are over 300 000 km of sewers, nearly 20 000 sewage-pumping stations, and at least 8000 separate sewage-treatment works operating continuously. A few of these treatment works are of considerable size and complexity, serving the equivalent of several million people in urban areas, while at the other extreme there are hundreds of small treatment works serving villages and hamlets throughout the country. In addition, there are thousands of privately owned sewage-treatment systems serving isolated individual properties and housing developments. Overall, sewage treatment in the UK involves the employment of about 30 000 people and a financial turnover approaching £3 billion per annum.

However, such a major highly developed and valuable asset, dedicated to the control of water pollution, did not materialize in just a few years. The development of our sewerage systems and sewage-treatment works to their current status stretches back to the middle of the last century. It is therefore appropriate in this book, to outline the history of sewage treatment before proceeding to describe the nature of sewage in some detail and the various processes which are used to treat it. In a final section, the control, administrative arrangements and training provisions for sewage treatment in the UK are reviewed.

2. HISTORICAL DEVELOPMENT

NINETEENTH CENTURY

During the early nineteenth century, the pollution of rivers in the UK became widespread, following the increase in piped water supplies to supply factories which were established during the industrial revolution.

The water-carriage system for sewage was employed by many of the ancient civilizations, and there are remains of several Roman examples. However, these fell into disrepair and most towns throughout the centuries lacked this basic amenity. It was not until 1810 that water closets were again being installed, and by about 1850 they had become increasingly popular in most parts of the country. The water closets originally discharged into closed pits but, to avoid the particularly unpleasant task of emptying them, the practice gradually developed of constructing overflows to the street rain-water drains. As these drains had not been designed to convey foul matter and were constructed merely as open gullies, it is easy to visualize the noxious conditions which resulted, and the filthy conditions under which the early nineteenth century town-dwellers lived. Indeed, the all-pervading stenches were too much – even for the hardened noses of those days, and the authorities attempted to impose penalties on those who used rain-water drains to carry sewage. The practice, however, had become too widespread to be stopped and penalties had little effect, with offenders often omitting the cesspits and connecting the water closets direct to the street drains.

Rivers and wells became polluted and in these conditions diseases and fevers were rampant, particularly those caused by waterborne organisms. For example, one epidemic of cholera in the UK was responsible for 50 000 deaths. To remove such dangerous and intolerable conditions, the next development for large cities was the construction of underground sewers to convey the sewage from houses to the nearest watercourse. Whilst this brought obvious benefits, it simply transferred the nuisance and hazard from the streets to the rivers and streams. The result was the immediate fouling and progressive deterioration of natural watercourses.

Industrial Revolution

The industrial revolution had a two-fold effect. Firstly, it caused a sudden concentration of population in the new centres of industrial development which resulted in the discharge of large quantities of offensive solids and liquids into short stretches of streams, whereas previously the polluting matter reaching the rivers from rural towns and villages had been smaller in quantity, was more widely distributed and relatively easily absorbed and purified. Secondly, industrial development led to the discharge (to streams) of increasing volumes of effluents from industrial processes. These significantly increased the fouling of watercourses, and in many cases their effects were even more serious than the pollution by domestic sewage. The following account of the experiences of a distinguished foreign visitor to London in the middle of the last century illustrates the condition of the River Thames:

'The cry of the day seemed to be "India is in revolt and the Thames stinks". It was hard to tell which circumstance was more disturbing to the British public, the mutiny in provinces far beyond the seas or the odours arising from the once proud river washing the shores of the Houses of Parliament. Here frightened legislators resorted to the use of curtains saturated with chloride of lime to mask the stenches wafted into the parliamentary chambers from the river, and travellers went far out of their way to avoid passing over the bridges spanning the foetid floods of the lower Thames.'

Similar conditions prevailed in other parts of the country, and one witness appearing before a Royal Commission on River Pollution Prevention, in 1868, confirmed his statements as to the outrageous foulness of the River Calder by writing a letter to the Commissioners using, instead of ink, "water" taken from the river, adding: "Could the odour only accompany this sheet also, it would add much to the interest of this memorandum".

Fish kills were severe and in one incident in the River Tame below Birmingham in 1867 about 60 000 fish were estimated to have died.

Rivers Pollution Prevention Act 1876

During the latter half of the nineteenth century, several Royal Commissions and Select Committees investigated methods for the treatment and disposal of sewage, and for the prevention or control of river pollution. A major result of these deliberations was the passing of the Rivers Pollution Prevention Act 1876. This Act, which remained in force until 1951, made it an offence to discharge solid wastes or any poisonous, noxious or polluting liquid into a river (although industry was protected to some extent). The enforcement of these provisions was in the hands of the local authorities, who became responsible for treating the sewage draining from their areas, and hence for providing and maintaining sewage-treatment works. Manufacturers were under a similar obligation to treat their industrial effluents before discharge to a stream unless, as became customary, they arranged for a local authority to accept them into a public sewer for treatment with the ordinary sewage. The researches of Pasteur and others into the nature and metabolism of bacteria gave a new dimension to the treatment of sewage and industrial effluents. Simple application to land was gradually superseded by more controlled methods of filtration such as intermittent downward filtration and fill-and-draw contact beds, until the biological filter (Chapter 7) was established as a reliable method of oxidation of the organic impurities in sewage and industrial effluents.

ROYAL COMMISSION ON SEWAGE DISPOSAL

The efforts of the previous fifty years culminated in the appointment, in 1898, of the Royal Commission on Sewage Disposal "to enquire and report what methods of treating and disposing of sewage (including any liquid from any factory or manufacturing process) may properly be adopted". The Commission sponsored a considerable amount of research, not only into treatment processes but also into analytical methods. In its ten reports to 1915, it laid the foundations for modern sewage treatment, and concluded that biological filtration, as still practised today, was superior to the earlier attempts at filtration. It studied the effects of sewage effluents in rivers, and set standards for these effluents in relation to the dilution available[1] (as described in Chapter 3).

The enforcement of the river pollution law was in the hands of the county councils and county borough councils until the formation of River Boards in 1948. These boards were responsible for pollution control in England and Wales, and continued until 1964 when they were replaced by the River Authorities with additional powers. Similar River Purification Boards for Scotland were formed two years later.

RECENT LEGISLATION

Relevant enactments of Parliament include the Public Health (Drainage of Trade Premises) Act 1937, Rivers (Prevention of Pollution) Acts 1951 and 1961, the Clean Rivers (Estuaries and Tidal Waters,) Act 1960, the Public Health Act 1961 (Part V) and the Water Resources Act 1963, all of which applied to England and Wales. The corresponding Acts for Scotland were the Rivers (Prevention of Pollution) (Scotland) Acts 1951 and 1965.

As a result of various reports, including 'Taken for Granted' in 1970[2], two further major pieces of legislation were introduced – the Water Act 1973 which set up ten regional water authorities in England and Wales (as described more fully in Chapter 12), and the comprehensive Control of Pollution Act 1974 which brought together previous legislation extending control to groundwater and coastal waters. Most of Part II of the Control of Pollution Act 1974, which contained the more effective regulatory elements, did not become effective until 1985.

A major change to the water industry in England and Wales occurred in 1989 when the water supply and sewage disposal functions of the ten regional water authorities were taken over by privatized Water Service Companies. At the same time the responsibility for the management of water resources and for pollution control of rivers, watercourses, estuaries and coastal waters up to the 3-mile limit, together with fisheries, recreation, conservation and navigation, was vested in a newly formed National Rivers Authority comprising ten regions approximating to the areas of the new Water Service Companies.

The National Rivers Authority function was separated from the privatized water companies so that the regulatory function for discharges from sewage-treatment works to controlled waters did not cause a conflict of interest. Concurrently, the national regulatory body the Office of Water Services (OFWAT) was formed, to promote efficiency in the industry and ensure that the interests of the public were protected, especially with regard to price control.

In 1996, the Government established the Environment Agency which combined the functions of Her Majesty's Inspectorate of Pollution, the National Rivers Authority and the Waste Regulation Authorities (which had been departments of the County Councils), thereby providing an integrated pollution control authority. In Scotland, the equivalent body is the Scottish Environment Protection Agency (SEPA), formed from Her Majesty's Industrial Pollution Inspectorate, the River Purification Boards, the waste regulation sections of the District Councils, and the local authority air pollution functions[3].

EUROPEAN UNION LEGISLATION

In addition to national legislation, requirements of the European Union (which are contained in Directives) have to be met for aspects such as the quality of river water to be abstracted for water supplies, rivers designated as fisheries, and the quality of bathing waters, which in this country are principally coastal beaches. A Directive designed to control the release of dangerous substances resulted in 'daughter' Directives covering the discharge of such substances as mercury and cadmium[4]. The parent Directive also refers to various organic substances including pesticides.

Urban Waste Water Treatment Directive

The main legislative drive for current and future investment in sewerage and sewage treatment is the Urban Waste Water Treatment Directive[5] which has, as its main objective, to ensure that significant discharges of sewage are treated before they are allowed into inland surface waters, groundwaters, estuaries or coastal waters. For the purpose of the Directive, significant discharges are those to fresh waters or estuaries serving communities with population equivalents of more than 2000, or those to coastal waters serving communities with population equivalents of more than 10 000. Sewage will usually be treated to secondary treatment standards (normally a biological process) but, in estuarine and coastal areas with high natural dispersion characteristics, primary treatment may be acceptable. Discharges into areas which are identified as 'sensitive' require more stringent treatment than the secondary treatment specified in the Directive. The Directive also requires appropriate treatment to be provided for discharges from smaller communities and an end to the disposal of sewage sludge at sea. The compliance dates given in the Directive range from 1998 to 2005, depending on the size of the community and location of the discharge. Separate provisions in the Directive refer to a number of industrial sectors, the discharges from which are of a similar nature to domestic sewage. The Urban Waste Water Treatment Directive is transposed into the Urban Waste Water Treatment (England and Wales) Regulations 1994[6].

CONSERVATION AND RE-USE

In nature, flora and fauna respond to changes in the environment by either modifying or moving; otherwise they may be overcome by the changes. Before the Industrial Revolution, many watercourses were harnessed for drinking, irrigation, power, waste disposal and processing. Industrial man has endeavoured to exploit and adapt the environment to support his needs, often resulting in significant long-term damage.

The manufacturing industries which developed during the Industrial Revolution had need for iron, coal, manpower, transport, customers and water. When these coincided, the watercourses became over-used and then heavily polluted. At first, the needs of the industry were accepted and the damage to land, water and the air was tolerated. The importance of employment was generally accepted as being more important than the health of rivers. To avoid penalizing industry, for many years the Public Health (Drainage of Trade Premises) Act 1937 gave prescriptive rights to industrial discharges. Established companies had these rights gradually eroded as their factory processes changed with time. The gradual erosion of prescriptive rights by successive laws reflects the difficult choice between the provision of jobs and protection of the environment.

The industrial need for water was satisfied by abstraction from rivers, estuaries and boreholes. As long as there was a plentiful supply of water there was little need for manufacturers to treat and directly re-use their industrial effluent. As charges for industrial effluent treatment increased, manufacturers were prompted to evaluate the costs and benefits of water re-use. In recent years the increasing cost of water supply and the reception and treatment of industrial effluents into the sewerage system and treatment works has encouraged many industrial users to pretreat their wastewater before discharge and, where possible, re-use processed water. The drought conditions which have been experienced in recent years have also concentrated the mind of the industrial water user as well as the water companies and Government regulators. The waste or leakage of water is also a key concern for water users. Because water supply and effluent disposal can represent a significant cost to an industry and hence its profitability, a reduction at source of water consumption, leakage, wastewater, pollutants and raw or processed materials, must be factors in determining the viability of an industrial process.

Where it has not been possible to eliminate the use of water, there are a number of methods of pretreating the wastewater for re-use or discharge, many of which are similar to domestic sewage-treatment processes. The re-use of cooling water, which often ran to waste, represents the simplest re-use of wastewater. In water-cooling systems where the water was allowed to pass through once and then to a sewer or watercourse, the introduction of air-cooled or fan-assisted water-cooling often provided a remedy. In some processes it is possible to collect cooling water and use this in other factory processes before discharge.

There are many consultancies and trade organizations which will advise on the reduction, elimination or re-use of wastewater. A discharge of any wastewater to a sewer represents a loss of that water from the nearby river or aquifer system. The vital importance of recycling high-quality treated effluent back into rivers is emphasized in those rivers where more than 50% of the flow originates from treated wastewaters. The higher incidence of drought in some areas and the gradual increase in water abstraction and use has placed a greater strain on fish and other aquatic life.

3. REASONS FOR WASTEWATER TREATMENT

Wastewater (or sewage) contains pollutant materials which are suspended or dissolved in a relatively large volume of water, together with silt, fat, grease, detergents, paper, fibres and many micro-organisms. If the untreated sewage was directly discharged to a watercourse it would cause severe pollution and present a danger to public health. Sewage-treatment works are designed to remove the gross polluting material and produce a treated sewage which is acceptable to the environment. An aerial view of a typical medium-sized sewage-treatment works is shown in Plate 1.

MEASUREMENT OF WATER QUALITY

Because treated sewage will inevitably arrive at a watercourse, estuary, sea or in groundwater, it is essential that the environmental impact is quickly and accurately measured and the short-term and long-term effects are predicted. The general characteristics of sewage are described in Chapter 4. The main reasons for sewage treatment are the elimination of waterborne diseases, prevention of pollution, or ecological damage to the receiving water and its environs with consequent loss of recreational benefits. In addition, the aesthetic impacts due to visual and smell nuisance are unacceptable. The legitimate use of water resources by water undertakers, agricultural and industrial abstractors also requires that sewage is suitably treated to minimize the impact upon receiving waters.

In order to decide on the method and extent of treatment required to produce an effluent which is suitable for discharge to the receiving water, and also to permit day-to-day control of the processes, certain standard analyses of the sewage must be carried out. The two principal analyses performed are the biochemical oxygen demand (BOD), which is a measure of the organic strength (and hence indicates the extent to which oxygen is consumed by the sewage or effluent), and suspended solids

Plate 1. Aerial view of a typical medium-sized sewage-treatment works

(SS), which is a guide to sludge production and possible siltation in the river. There are many other tests for chemical constituents such as chloride, pH, suspended solids, ammonia and nitrate. Fortunately, the empirical test for biochemical oxygen demand (BOD) has evolved as an arbitrary but comparative test procedure designed to quantify the effect of a sewage discharge on a watercourse. Because (a) the BOD test is a standardized procedure and is incorporated into statute[6] and many legal documents, which bind the discharger and the Environment Agency, and (b) there is no satisfactory alternative, the test is a keystone in the assessment of the strength and impact of a wastewater.

Biochemical Oxygen Demand

The BOD test measures the amount of dissolved oxygen absorbed by sewage under standard conditions of temperature, time and analytical procedure. Whilst the accuracy is only about $+10\%$ and with a lower detectable limit of about 1 mg/l, it is successfully accepted throughout the world because it simulates the natural biological processes in a watercourse. The BOD is a measure of the absorption of dissolved oxygen by the biological, biochemical and some chemical processes of the organisms and materials in the sample during a five-day period at a standard temperature of 20°C.

If the BOD test was determined for about 30 days, the result would approach a limiting value, reflecting the almost complete biochemical oxidation of the biodegradable pollutants in the sample.

The BOD load absorbs the dissolved oxygen from the watercourse as it is oxidized to form more stable compounds and hence reduces the dissolved oxygen which is available for fish and other aquatic life. A high BOD, such as that from some industrial effluents, may remove the dissolved oxygen for many kilometres of the watercourse, and, if prolonged, may destroy the 'life' of the watercourse and therefore its ability to self-purify or accept small increments of pollution load.

The analytical results enable sewage-treatment works to be designed and operated to minimize the effect upon the receiving water. There are other analytical tests (such as the permanganate value) which give a rapid measure of the polluting properties of a sewage and are partly related to the BOD.

Chemical Oxygen Demand

The chemical oxygen demand (COD) test is carried out using a boiling solution of strong acid plus an oxidizing agent and represents a much higher degree of oxidation of organic matter than the BOD test. There is generally a relationship between the COD and BOD of a particular sewage. Because the COD test is easily automated and its accuracy is easily reproduced, it is often used by water plcs to calculate the additional industrial effluent load on a domestic sewage-treatment works – enabling a charge to be calculated which reflects the capital and operating costs of treatment and disposal. Whereas most of the BOD is removed in the treatment process together with significant amounts of the COD, the biological process may not be able to oxidize or metabolize part of the COD during the retention period in the sewage-treatment processes, and this COD will pass out with the treated effluent and may place a so-called 'hard COD' load on the watercourse.

Suspended Solids

Suspended solids may be a mixture of inert solids or live and dead organic matter, thus they may also contribute to the BOD load. The suspended-solids test measures the solids which readily settle as well as many particles which may remain in suspension for some time. Any very fine particles which are not retained by a 0.45 μm pore diameter filter are not included in the suspended-solids test. The test is carried out by filtration or centrifugation and the deposited washed solids are dried at 105°C for one hour.

In its Eighth Report in 1912, the Royal Commission introduced the innovative concept of dilution ratios for secondary-treated sewage effluent discharging to watercourses. The Royal Commission standard, i.e. 20 mg/l BOD and 30 mg/l SS with a minimum dilution ratio of 8:1, set the pattern for river-basin management.

EFFECTS ON AQUATIC LIFE

The suspended solids may blanket the stream bed and prevent respiration of the benthic fauna and flora which depend upon a free interchange of dissolved oxygen between the water and 'bottom layers'. Suspended solids may also blanket the flora and prevent photosynthesis as well as interfere with the respiration of fish and other aquatic life.

There are many constituents of domestic sewage and industrial effluents which may have an effect upon the life in a watercourse. The breakdown of protein, urea, and other nitrogenous compounds will release ammonia, which is potentially toxic to fish at concentrations as low as 0.01 mg/l. When ammonia is present with other pollutants, the additional synergistic effect may greatly magnify the toxicity; this is especially acute when coupled with low dissolved oxygen and increased temperature.

The common elements nitrogen and phosphorus, when present as the nutrients nitrate and phosphate and discharged from sewage-treatment works or as farmland runoff, may cause an enrichment of the nutrient balance and encourage algal bloom and excessive plant growth, and may also have a direct toxic effect upon flora and fauna. Concentrations as low as a few micrograms per litre of some heavy metals can have a pronounced effect upon aquatic organisms.

4. DOMESTIC AND INDUSTRIAL EFFLUENTS

GENERAL CHARACTERISTICS

Domestic sewage includes human wastes and all other waterborne wastes from domestic properties. In many older sewerage systems, surface waters and drainage from streets and roofs may enter the sewers which are therefore called 'combined sewers'. The result is that during heavy rain the volume of sewage may be many times that in dry weather. In more modern systems the surface runoff from rain is usually collected in entirely separate sewers which discharge direct to a stream or river and the volume of sewage flowing in these foul-water sewers should, in general, be less affected by rainfall. As a compromise, a partially separate system of sewers may be used, in which some house-roof drainage could conveniently be allowed into the foul sewer, resulting in flows which fall between these two extremes. A certain amount of groundwater, or 'infiltration', also finds its way into most sewerage systems and, with older or damaged sewers, leakage and consequent structural damage (as well as pollution) may occur.

Industrial effluents may also be present in greatly varying proportions. The industrial component may either be contributed by one type of manufacturing operation or may consist of wastewater of widely varying composition from many different industries; indeed at some treatment works the volume and total strength of the industrial effluent may greatly exceed that of the domestic sewage.

Industrial effluents may contain toxic metals (such as cadmium, chromium, copper, tin, nickel, lead, zinc), sulphides, rapidly biodegradable organic matter, complex biochemicals from pharmaceutical waste, waste foodstuff, oils and greases. Untreated industrial effluents may (a) attack the fabric of the sewer due to extremes of pH, (b) produce or cause evolution of toxic gases, (c) overload or inhibit the physical or biological processes at the treatment works, or (d) pass through the works to inflict further damage on the receiving watercourse. There is a comprehensive regulatory system operated by the Water Service Companies, the Environment Agency and the Scottish Environment Protection Agency, supported by legal sanctions and claims for damages.

The Water Service Companies levy a charge for the reception, treatment and disposal of industrial effluents reflecting the total cost for disposal. Because of the varying degrees of load, treatment effects, toxicity and inhibiting effects, formulae have been developed to reflect the true cost and difficulty of sewage treatment and disposal[7].

The mixed sewage conveyed to treatment is therefore complex. Not only is its character influenced by the type and proportion of industrial effluent present, other factors can also have a considerable effect; for example (a) the construction and hydraulic characteristics of the sewers, (b) the length of the sewers and the time taken in travelling to the inlet of the treatment works, (c) whether the sewage has been pumped, and (d) the hardness of the carrying water. No two sewages are alike: some are much 'stronger' or more concentrated, and others are more resistant to treatment. The volume of sewage per head of population also varies from place to place, depending on the type of development, the amount of industrial effluent included, and the extent to which rainfall enters the system. The dry-weather flow at a typical treatment works may vary up to about 200 litres per person per day where the industrial contribution is small. The flow and the strength also vary with the time of day, day of the week and the activities of industry.

Although sewage-treatment works conform to the same broad principles, they must be designed and operated in accordance with the characteristics of the particular sewage to be treated and the quality of the effluent required.

The impurities are mainly organic in origin, associated with much inorganic matter, and may be divided into three categories; (a) substances suspended in the liquid, (b) substances dissolved in the liquid, and (c) finely divided 'colloidal' substances midway between these two. A suitable analogy might be: a cup of tea in which there are tea leaves suspended in the liquid, sugar dissolved in it, and the constituents of milk in a finely dispersed 'colloidal' form, giving the liquid its creamy appearance.

TREATMENT PROCESSES

The sewage needs treatment to a standard set by the Environment Agency or the Scottish Environmental Protection Agency according to the nature of the receiving water. There are many chemical, physical and biological processes which could be used to treat sewage to an acceptable standard for discharge to rivers, lakes, estuaries and seas – all with varying degrees of success, technology and cost. The presently preferred methods usually include screening to remove large particles, settlement to remove the heavy gritty or inorganic matter, followed by further settlement to remove most of the remaining solid impurities from the liquid portion. The resultant clearer liquid, which contains dissolved and colloidal matter, is then subjected to a biological stage where most of the remaining impurities are oxidized to carbon dioxide and water, or converted into biomass which is removed by final settlement before discharge to the receiving water.

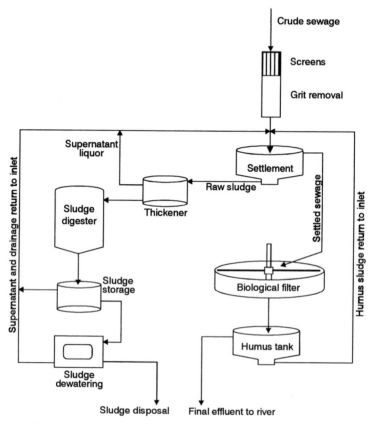

Fig. 2. Basic layout of a medium-sized sewage-treatment plant with biological filtration

The suspended matter which settles in the sedimentation tanks retains from ten to thirty times its weight of water, and forms a bulky deposit of wet sludge. There are many sludge treatment processes which remove liquid from the sludge, thus concentrating and stabilizing the solids so that they may be safely and economically disposed of. Because this separation is incomplete, leaving a liquid portion still containing matter in solution and suspension as well as the sludge portion still associated with a considerable quantity of liquid, the treatment of sewage is complicated. In addition, ammonia is present, derived from body wastes and some industrial effluents. Inorganic ammonia is soluble and not removed by settlement, but it can be oxidized to nitrate by two types of bacteria acting in sequence in the presence of oxygen. This process of 'nitrification' occurs naturally in soils and rivers and can be specially designed into the treatment of sewage to remove unacceptable concentrations of ammonia. Ammonia, even in very low concentrations, may affect fish life in rivers and cause additional problems in treating river water for public supply.

Ammonia, nitrate and phosphate are present in crude and treated domestic sewage and contribute towards the eutrophication of watercourses, resulting in increased algal and weed growth with the consequent problems of de-oxygenation, overload and restrictions in navigation and leisure uses.

A simplified diagram showing the basic layout of a sewage-treatment plant is shown in Fig. 2.

5. PRELIMINARY TREATMENT

The settlement of sewage to separate the suspended solids is usually preceded by two stages of preliminary treatment (termed 'screening' and 'grit separation') to remove large solid objects and the grit which is derived mainly from road and roof surfaces.

SCREENING

At most works it has been customary to pass the sewage through screens made of bars spaced up to 25 mm apart, so that any suspended or floating material larger than this is deposited on the screen. A coarser screen with about 100-mm spacings often precedes the main screen. The collected material is normally moved by mechanical rakes. However, the use of 6-mm fine screens (Fig. 3) is most often the first choice, because they remove greater amounts of material and can have aesthetic and hygienic benefits.

The primary objective of screens is to intercept rubbish, wood, large rags, etc., which would clog channels and pipes, damage pumps or pass through the treatment processes and be discharged to the watercourse. Screens also arrest a considerable amount of smaller material, such as paper, peelings and faecal matter. The quantity of screenings depends largely on the characteristics of the sewerage system and the size of the screen apertures. The distance that material of this type has to travel in the sewers and the greater the turbulence, the more likely it is that these solids will be disintegrated and hence not be intercepted by the screens. Certain industrial effluents, e.g. from the textile industry, can also contribute considerably to the total quantity of screenings.

Manufacturers have responded to the need for improved screening methods by developing a wide variety of machinery to reduce the problems of screenings throughout the works and beyond.

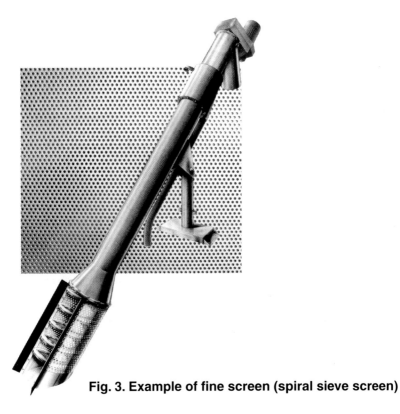

Fig. 3. Example of fine screen (spiral sieve screen)

Screenings are obnoxious and their hygienic disposal can be difficult. Three methods of disposal which have been employed are (i) burial, often after washing and compression into bales or bags, (ii) incineration using oil or spare gas produced at a later stage in sludge treatment, or (iii) maceration into very small particles and return to the sewage for later treatment in conjunction with the main bulk of the sludge.

Comminutors, consisting of a hollow drum with slots and teeth which rotate in the flow and engage with steel combs, have generally been replaced by improved disintegrators located in an open channel or as part of a pumping unit so that the screenings are effectively broken up and may be co-settled with the raw sludge in the primary sedimentation tanks.

GRIT SEPARATION

It is important to remove the relatively heavy inorganic grit and silt before the main sedimentation process because it is liable to damage mechanical equipment such as pumps and valves or, if allowed to pass into the primary sedimentation tanks, to form semi-solid banks of grit which are difficult to remove. Sludge mains, digesters and storage tanks, may also become choked and sludge treatment made more difficult. The removal of this grit and silt, when intercepted in relatively small tanks or channels designed for the purpose, easily lends itself to mechanization.

The function of grit channels is to selectively deposit the heavy inorganic material. If a velocity of 0.3 m/s is maintained, the grit settles whilst the lighter organic matter is carried forward, so that only reasonably 'clean' grit is deposited. In some cases the grit may require washing to remove the remaining organic matter, prior to disposal.

Grit channels or tanks consist of catch-pits or widened channels of special cross-section shape, placed between the screens and primary sedimentation tanks and designed or controlled by a downstream control flume, to maintain a velocity of 0.3 m/s. As the flow varies widely with weather conditions and time of day, the development of a design in which this velocity is automatically maintained under all conditions is not easy. Typical designs of grit interceptors include spiral-flow aerated channels and vortex-type as well as cross-flow square tank types.

On larger works the deposited grit is normally removed from the collecting chamber by rakes, bucket dredgers, grabs, pumps or air lifts. Mechanisms without working parts permanently under water, and hence not severely abraded by grit, have much to commend them.

The quantity of grit collected varies widely, depending upon the type of sewerage system, climatic conditions, and (in the case of combined sewers) the nature of the area drained and of the street surfaces.

Washed grit has been used for filling hollows, making embankments and site roads. However, because of needles, razors, sharp glass and metal, coupled with potential rat infestation, great care must be taken when handling the material.

STORM-SEWAGE TREATMENT

During storms, a works may receive flows greatly in excess of six times dry-weather flow, depending upon the degree of separation of rainwater from the sewage flow and the age and condition of the sewerage system. The high velocity of the sewage scours the deposited sediments in the sewers and hence presents a very high volumetric, solids and organic loading to the works. To prevent overloading of the works and flushing out the treatment biomass, as in the activated-sludge process, the Environment Agency accept that it would be prohibitively costly to design works with the capacity to treat these initially very strong storm flows as part of the continuous flow-through treatment of normal sewage. The generally accepted treatment for storm sewage is based on designing a sewage-treatment works to provide full treatment for up to three times the dry-weather flow (DWF), which also allows for appreciable diurnal variations. The first flush of flows between 3 DWF and 6 DWF is often considerably stronger than normal sewage and needs to be diverted to storm tanks which, like sedimentation tanks, allow sufficient time for settlement. When full, they allow a significantly weaker liquid to be discharged to the watercourse, which by this time is already swollen with normal rainfall runoff. When the storm has abated, the settled storm sewage is returned to the works for treatment, and the settled sludge may be pumped to sludge storage or thickening tanks. Storm flows in excess of 6 DWF are usually allowed to discharge direct to a watercourse with the consent of the Environment Agency, as they are considerably diluted by the rainfall and the high river flow.

6. PRIMARY TREATMENT

SEDIMENTATION

Following grit separation and screening, sewage is treated by settlement in large tanks. In this process, suspended solids gradually settle and are removed as a liquid sludge. The sludge has a dry-solids content of about 5%, of which 70-80% is organic and volatile matter. The organic matter includes fats, grease, food residues, faeces, paper and detergents, whilst the inorganic matter mainly consists of siliceous grit. Although raw sludge represents only 1-2% of the total flow of sewage, the treatment and disposal is so capital and labour intensive that it can amount to 50% of the total cost of running a works[8].

Sedimentation is accomplished by passing the sewage through tanks slowly enough for the suspended particles to drop to the floor of the tank, whilst the settled liquid passes forward to the biological treatment process. The tanks which are used for sedimentation are generally of the rectangular horizontal-flow, upward-flow or circular radial-flow type. Most new tanks are radial flow because of design and operational benefits.

Horizontal-Flow Tanks

These are usually rectangular in plan, 1.8-3.0 m deep, and their combined capacity may be equivalent to 8-12 hours' flow. They are usually built slightly deeper at the inlet end, where most of the sludge is deposited. Sewage enters and flows horizontally to an outlet weir at the opposite end. The deposited sludge accumulates on the floor and is scraped, underwater, to sludge hoppers and outlets where it is withdrawn at regular intervals.

Upward-Flow Tanks

These are either circular or square in plan. Sewage enters at the centre of the tank and at as low a level as possible, consistent with causing no disturbance of the settled sludge. It then passes slowly upwards and outwards to overflow weirs around the periphery, the settled sludge being left behind. The bottom of the tank, in which the sludge collects, is in the shape of an inverted cone or pyramid. A discharge pipe, rising from this sludge hopper and terminating in a chamber at a lower level than the water surface, enables the liquid sludge to be forced up hydrostatically when the valve is opened. This type of tank is more costly to construct than the rectangular horizontal-flow tank due to its depth so that, while many remain in use, they are usually only built for small package plants.

Radial-Flow or Mixed-Flow Tanks

The disadvantages of upward-flow tanks led to the introduction of radial-flow or mixed-flow tanks (Plate 2). They are normally circular in plan, with the floor gently sloping towards the centre. Sewage enters at the centre below the water level and flows outwards and partly upwards to an outlet weir around the circumference. The deposited sludge is slowly moved by a mechanical scraper across the tank floor towards a central sludge sump, from which it is withdrawn at intervals by manual or automatic means. The increase in site automation and computer control, coupled with reliable automatic sludge pumping equipment and control, has resulted in a move towards automatic desludging.

Plate 2. Radial-flow tank showing arrangement of sludge scrapers

Horizontal-flow and radial-flow tanks usually have devices for removing unsightly scum which would collect on the surface and block the sparge holes of biological filters. The radial-flow tank is the preferred design because it enables effective and automatic scraping and sludge removal, which benefits the subsequent stages of sludge treatment.

Effective sedimentation removes the bulk of the suspended impurities (50-70%) but only removes 20-30% of the BOD from the sewage. The resulting tank effluent is still turbid and contains all the dissolved impurities and colloidal matter. It is also still putrescible, and further treatment is required to significantly reduce its impact on a watercourse.

7. SECONDARY TREATMENT I

The dissolved and colloidal impurities, which remain following the primary treatment process, are subjected to a biological oxidation stage in which up to 80% of the remaining BOD is removed and converted to carbon dioxide and a secondary sludge (humus) which is separated from the final effluent before it is discharged to the watercourse. The biological oxidation may be achieved in two ways; various forms of 'captive' film *biological filters* and the *activated-sludge process* in which there is a continuous agitation and aeration of the biological growth (Chapter 8).

BIOLOGICAL FILTRATION

Many years ago, further purification of sewage was originally achieved by applying the liquid to land. This process is rarely used nowadays because of the large area of land required. Whether the 'broad irrigation' or 'intermittent downward filtration' system was used, only a portion of the land was in use at any one time – the remainder having to be rested to re-aerate the soil and to allow the deposited sludge to be assimilated into the land. The areas of resting land were often cropped, hence the term 'sewage-farm'.

In land treatment, the purifying agencies are the bacteria and other forms of life naturally present in the soil, which in the presence of air metabolize the impurities with which they come into contact as they would with manure or dead vegetation. Under suitable conditions the effluent is clear and sparkling, stable and inoffensive. The introduction and development of biological filters to exploit the micro-organisms produced a more efficient process and improved the process control.

Filter Construction and Operation

The modern biological filter consists of a bed of solid medium 1.5-2.0 m deep, over which the settled sewage is distributed. The medium is generally broken stone or slag about 50 mm in size, although some plants use different sizes. Plastic media in the form of corrugated sheets or random-packed stone-sized pieces are also used. They have the advantage of a large ratio of surface area to mass, and can be used in deeper beds than are possible with stone media, with economy in both area and structural costs. The surface of the medium becomes covered with a jelly-like film in which bacteria live together with protozoa, fungi, insect larvae and adult flies. As the settled sewage percolates downwards over these biologically active surfaces the microscopic organisms use the sewage impurities as nutrients for their growth and reproduction. Free access of air into the bed is vital to the process because the organisms are aerobes, therefore needing oxygen. The varied animal life plays an important part in the process by feeding on the accumulating bacteria and fungi, thus reducing blockage by excessive film growth. The word 'filter' is a misnomer because it is not designed to physically filter out suspended solids; its function is to present an extended surface on which the necessary bacteria and the sewage can be brought into intimate contact in the presence of air and metabolize the organic impurities.

Various methods are used for distributing settled sewage over the surface of a filter. The most common type of distributor for circular filters (Plate 3) consists of radial revolving arms suspended from a central column; such installations have become the layman's recognized identification of a sewage-treatment works. These 'sprinklers' revolve due to the reaction of the jets of settled sewage issuing from the holes in the horizontal arms. The speed of rotation may be controlled to achieve maximum

Plate 3. Circular biological filters

efficiency and, when the head or pressure of sewage is insufficient, electric motors may power the distributors. When rectangular filters are installed, a travelling distributor spans the bed and moves backwards and forwards longitudinally. The motive power may be due to the flow, but new installations normally incorporate electric drives.

The 'loading' which can be applied to a biological filter depends upon (a) the amount and type of impurity left in the sewage after sedimentation, (b) the condition of the respiring biomass, and (c) the characteristics of the filter medium. The effluent draining from a properly operated bed is almost clear, but always contains suspended matter derived mainly from the breaking away of particles of active film from the filter medium. The quantity of this suspended matter is subject to seasonal fluctuations and must be removed before the effluent is discharged to a watercourse. For this purpose 'humus tanks' are employed which are similar in design to primary sedimentation tanks. The final effluent can then be passed to a watercourse, whilst the intercepted solid matter (known as humus sludge) is dealt with by one of the various methods of sludge treatment described in Chapter 10.

Biological filters, followed by humus tanks, should produce a comparatively clear and inoffensive effluent which will often satisfy the requirements for discharge to a river for both organic impurity removal and ammonia removal by nitrification. They also provide a robust and reliable method of treatment, usually without the need for electricity. However, (i) the area of land required is large and at least five times the area needed for an activated-sludge plant, (ii) filters form a favourable habitat for various species of small flies which sometimes cause nuisance near the works, and (iii) the final traces of suspended solids are difficult to remove and high-quality effluents cannot be achieved without further treatment. The development of intensified methods of using biological filters (dealt with below), has enabled the first two of these disadvantages to be reduced. The third can be overcome by employing one of the tertiary treatment methods described in Chapter 9.

ALTERNATIVE METHODS OF OPERATION

Recirculation

In this process (Fig. 4), the settled sewage (to be fed to filters) is first diluted with settled treated effluent which has passed through them before. The volume of recirculated effluent is set to about equal that of the incoming settled sewage, and the addition of this treated effluent significantly improves filter performance. It is understood that the improved performance is principally due to the wetting of a higher percentage of the media surface and increased hydraulic scouring which controls the slime accumula-

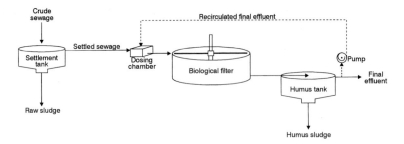

Fig. 4. Diagrammatic layout of a biological-filter plant with recirculation of final effluent

tion. About twice the load can be treated per unit volume of medium, and the actual overall rate of application of liquid (including recirculated effluent) is several times that in a conventional filter. At these higher loadings, without recirculation, slimy growths would soon fill the spaces between the separate pieces of medium and choke the bed, but through biological and physical agencies the recirculated treated effluent prevents their undue accumulation.

To achieve these advantages, larger feed pipes and distributors are required and larger humus tanks may be needed. Pumping is necessary to recirculate the treated effluent, but the associated additional capital and operational costs may be justified by the saving in filter-plant capacity.

The annoyance from filter flies is much reduced when recirculation is used.

Alternating Double Filtration

In this process settled sewage is passed through two filters in series; first through a primary filter and then through a secondary filter and at a fairly high flow-rate. The order of the two filters is alternated at intervals varying from one to seven days, the secondary filter then becoming the primary unit and the primary filter the secondary unit. Without alternation the primary filter will often operate as a high-rate filter, but in the alternating system both filters will have normal-graded medium and the growth of biological film, which forms in the primary filter, is loosened and washed out when this filter becomes the secondary unit. The overall growth of this film is thereby limited, and for this reason the system can deal with a heavier loading than ordinary filters, the benefits being similar to those of recirculation.

Alternating double filtration is particularily suitable for the treatment of strong carbohydrate wastes such as dairy wastes.

Combined Processes

Biological filters can be combined with other processes. In one combination the sewage is given a short period of treatment by an activated-sludge system (see Chapter 8) followed by biological filters. The dose rate to the filter can be increased considerably, and this technique has been successfully adopted in cases of overloaded filters or in dealing with particularly strong sewage.

OTHER 'CAPTIVE-FILM' SYSTEMS

Although not strictly biological filters as described above, a number of newer developments depend upon the growth of biological films on solid surfaces. The biological filter is the most common 'captive-film' system. 'Captive' or 'fixed film' refers to the attachment of the microbial film to an inert physical medium. This contrasts with the activated-sludge process (Chapter 8) in which the micro-organisms are kept in free suspension. Although most of the biological treatment processes have been used for many years, research work is continually being carried out to improve the performance and efficiency for environmental and commercial advantage. 'Captive-film' systems may employ intermittent submersion as in the rotating biological contactor or continuous submersion as in submerged-bed filters or fluidized beds[9].

Plate 4. Rotating biological contactor

Rotating Biological Contactors (RBCs)

These normally consist of closely spaced discs 2-4 m in diameter and a few millimetres thick, mounted on a horizontal shaft above a tank (Plate 4). The discs are made of lightweight plastic, metal or plastic mesh, although in some designs the discs are replaced with plastic pieces inside a cage. The discs form the medium on which the micro-organisms grow.

The unit is arranged so that about one third of the disc diameter dips into the sewage passing through the tank. As the discs slowly rotate, the organisms in the slime are alternately brought into contact with sewage and the atmosphere, so that treatment by biological oxidation takes place. The rotation also agitates the sewage in the tank, maintaining colonies of organisms in free suspension. Some designs provide for settlement before and after the biological oxidation stage in the same unit, whereas larger plants may rely on separate settlement tanks. Rotating biological contactors are compact plants, suitable for less than 100 people or, using larger units, may be combined for a population of several thousands. Because they are easily enclosed by an unobtrusive outer casing, the units can be partly installed below ground and generally require little attention. Rotating contactors are widely used for small communities, isolated hotels and institutions.

Fluidized Beds

These processes may be used in aerobic or anaerobic operation and they are in continual development. The biological growth takes place on sand particles kept in suspension by an upflow of settled liquor. It is necessary to ensure that sufficient oxygen is dissolved in the flow to enable aerobic treatment to take place, and excess growth must be separated from the sand. The sand particles provide a large surface area per unit volume, so that in an aerobic fluidized-bed application a high biomass concentration of 15-20 kg/m^3 is sustained and can achieve full nitrification in about 30 minutes. The process has also been used without free dissolved oxygen for the removal of nitrate from effluents in which the organisms then use the oxygen of the nitrate and liberate the nitrogen as bubbles of gas.

8. SECONDARY TREATMENT II

ACTIVATED-SLUDGE PROCESS

In the biological filter, the purifying agencies are micro-organisms living in a gelatinous film on the filter medium; the bacteria and other organisms are 'stationary' and only the sewage is in motion. In the activated-sludge process the micro-organisms are added to the incoming sewage and the mixture is continuously aerated and agitated in aeration tanks or channels so that the bacteria and sewage are intimately mixed. This bacterial culture is known as 'activated-sludge', consisting of flocculent clumps of bacteria and other microbes, termed 'flocs', in suspension. Activated sludge is initially developed by the prolonged aeration of sewage. Aeration for 6-10 hours with sufficient activated sludge followed by settlement of the sludge, usually in radial-flow tanks (as described in Chapter 6) produces a biologically stable, clear and well-purified effluent. Most of the activated sludge is then pumped back to the aeration tanks as a well-oxidized bacterial floc. The amount of floc increases as the impurities are removed from the sewage by growth of the sludge organisms; therefore a proportion is regularly or continuously drawn off and usually separately thickened at many large plants or passed to the primary sedimentation tanks for co-settlement with the raw sludge and thence to the sludge treatment process. The 'surplus' sludge is similar to the humus sludge removed after the biological filtration process. A simplified diagrammatic layout of the process is shown in Fig. 5. About 50% of the domestic sewage in the UK is biologically oxidized by the activated-sludge process.

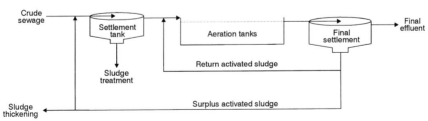

Fig. 5. Diagrammatic layout of an activated-sludge plant

Aeration Systems

Several methods of aeration are in use, divisible into two main categories, according to whether aeration is effected by diffused-air bubbles or surface-aeration devices.

Diffused-Air System. This system (Plate 5) uses compressed air which enters the aeration tank in the form of fine or coarse bubbles through 'diffusers'. The diffusers may be porous domes or finely perforated elastic membranes laid out on the tank floor. The air bubbles ascend and eventually burst at the surface, fulfilling the dual purpose of providing oxygen and agitating the contents to keep the activated-sludge particles in intimate contact with the sewage.

Plate 5. Diffused-air activated-sludge plant

Mechanical Surface-Aeration Systems. These rely on agitation at the surface of tanks – the constantly changing surface causing oxygen to pass into solution from the atmosphere. The agitation simultaneously keeps the sewage and activated sludge in intimate contact. Surface aerators may be mounted vertically or horizontally. With the vertical aerators (Plate 6), the aeration tank is usually divided into a number of compartments in the centre of which aerators, consisting of rotating paddles mounted on a vertical shaft, are fixed. These may include a vertical tube up which sewage and activated sludge is drawn and scattered over the surface, although similar circulation can be achieved without tubes. In the horizontal 'rotor' system, revolving longitudinal aerators with paddles or 'brushes' rotating on a horizontal shaft mounted just above the surface of the liquid, agitate the surface of the liquid – achieving the desired effect. In all these plants significant power is required, but this is offset by the much smaller area of land which is needed when compared with the usually gravity-driven biological filters.

Plate 6. Vertically mounted surface aerator

Nitrogen in the Activated-Sludge Process

As the activated-sludge process is a biological oxidation process, the oxidation products originate from the main 'organic elements' to form carbon dioxide, sulphates, nitrites, nitrates, phosphates etc., all of which are preferable to the complex and unstable pollutants in the crude sewage. Whilst the presence (in the final effluent) of nitrate is preferable to urea, ammoniacal or other nitrogenous compounds, it may cause excessive growths of plants and algae in rivers, and can present water-treatment problems where water is used for potable supply. High levels of nitrate in water derived from private wells (as opposed to the public supply) have been related to infantile methaemoglobinaemia ('blue baby' syndrome)[10]. Under some circumstances, the nitrate is readily broken down by facultative anaerobic denitrifying bacteria which use the nitrate as a source of oxygen. The resultant nitrogen gas can cause rising sludge in the final settlement tank and a deterioration of the effluent quality.

The denitrifying bacteria can be encouraged to reduce the nitrate before it reaches the final settling tanks by providing an 'anoxic' zone in the aeration tanks by careful control of the aeration and mixing conditions. The denitrifying bacteria need a source of carbon and this is usually supplied by the sewage.

Apart from the benefits of reduced nitrate to the watercourse, there is a reduced chance of rising sludge in the final settlement tanks, and improved settleability of the activated sludge[11].

ALTERNATIVE METHODS OF OPERATION

Deep-Shaft Process

The deep-shaft process differs from conventional diffused-air activated sludge in that a vertical shaft 30-150 m deep and 0.5-6.0 m diameter is used. The steel or concrete shaft may have a dividing wall for almost the full depth of the shaft to form 'downcomer' and 'riser' sections. Start-up air is introduced part way down the 'riser' and starts an air-lift action, drawing down the preliminary-treated sewage in the 'downcomer', so that a circulatory action is created. Compressed air is then introduced into the 'downcomer' where it is drawn down, aerating the sewage in admixture with the returned activated sludge, so that the air, firstly at increasing pressure and on the return towards the surface at decreasing pressure, has a 3-5 minute contact time compared to 15 seconds in a conventional activated-sludge plant. This physically and biochemically concentrated process requires a retention period of about 1 hour. When the process air has established circulation and a continuous path up the 'riser', the start-up air is discontinued. Because of the small footprint and high rate of treatment, the deep-shaft system is particularily suitable for domestic sewage and some high-strength industrial effluents. Whilst there are only about three deep-shaft processes in operation in the UK, there are many others treating sewage in Europe, USA and Japan.

Oxygen Activated Sludge

The low solubility of oxygen in water (11.3 mg/l at 10°C) limits the biological rate of respiration of the activated sludge. The existence of cold-blooded life in the rarefied oxygen concentrations of river and sea water indicates their susceptibility to even small biochemical oxygen demands from sewage. The rate of transfer of oxygen from the atmosphere through the boundary layer of the air-water interface and thence to the respiring cells, limits the respiration and the metabolism of waterborne organisms. Therefore the rate at which activated sludge removes pollution from sewage and the viability of the organisms is very sensitive to the amount of oxygen which is present.

The solubility of oxygen in water is also a function of the partial pressure of the complementary 79% nitrogen present in the air. Consequently any attempt to increase the oxygen content by using pure oxygen will significantly increase the amount of activated sludge which the process tank can sustain.

Process oxygen can be delivered to site as liquid oxygen at low temperature or produced on site by the 'molecular-sieve' technique using 'pressure swing absorption' (PSA).

In the 'Unox' process (which is a closed system), covered tanks are used to contain the oxygen above the activated sludge, dissolution being achieved by mixers; whereas in the 'Vitox' process (which is an open system) the oxygen is injected into a flow of activated sludge drawn by a pump from the main tank. High-purity oxygen processes have found particular application in the treatment of strong organic industrial effluents alone, or in admixture with sewage.

The choice between conventional activated sludge and the addition of oxygen depends upon the nature of the sewage, the works, and other local factors. The use of oxygen presents a number of advantages, including the production of a better-quality effluent with better sludge settling and dewatering properties[11]. Oxygen assistance may be of particular use where a plant is subject to shock loads, seasonal increases, or needs temporary improvement prior to new works.

Because of the dynamic nature of the activated-sludge process and the energy, air and varied sludge concentrations and inputs, the process (when compared with biological filtration) requires a greater understanding of the fundamentals for the optimum operating conditions to be achieved and operational problems to be anticipated, understood and overcome.

Extended Aeration

For the treatment of sewage from small or remote communities, easily installed prefabricated, extended-aeration plants are sometimes used. These may dispense with primary settlement and the sewage, after screening or disintegration, is admitted direct into the aeration compartment – giving a retention period of at least 24 hours. The mixed liquor then flows to a settling compartment, which is separated from the aeration compartment by a wall or baffle, and the activated sludge is returned by gravity or simple pump.

The loading of these plants is relatively low and the gradual increase in mixed-liquor suspended solids may be reduced by the concurrent breakdown of the older activated-sludge biomass and by more vigorous new growth. Surplus activated sludge must be removed at regular intervals otherwise solids will be discharged with the effluent at high flows.

Oxidation Ditch

Oxidation ditches are another form of the extended-aeration process in which the screened sewage is circulated with the activated sludge in a continuous loop channel. The concept exists in numerous variations, some of which provide settlement zones within the ditch, thereby eliminating the need for a separate settling tank. The essential thorough mixing and air entrainment are provided by a variety of horizontal rotating cages, brushes or rotors, and the rate of oxygen input is varied by adjusting the outlet weir and hence the depth of aerator immersion. Plants may serve populations of 1000 - 10 000. Because of their simplicity of design, they are of relatively low capital cost and can be constructed quickly.

COMPARISON WITH BIOLOGICAL FILTRATION

When compared with biological filters, activated-sludge installations (i) require less land, (ii) are cheaper to construct, (iii) need much less 'head' of sewage, and (iv) are not subject to nuisance from flies. However, because of the heavy power demand to drive air blowers or aerating mechanisms, they are more costly to operate. Some authorities consider that, from a chemistry and biochemistry point of view, activated-sludge plants are less robust than filters, are more easily upset by certain industrial effluents, and need skilled monitoring and supervision. They also produce relatively large quantities of thin watery sludge, which adds to sludge treatment difficulties. Because of the wide capital, operational and technical differences between activated-sludge plants and biological filters, the designer must carefully examine the options before choosing the secondary treatment method which best fits the local circumstances.

The activated-sludge process is a good example of a biochemical process in which a large biochemically reactive mass is controlled to achieve satisfactory performance parameters. The system variables of dissolved oxygen, power input, mixed-liquor suspended-solids content, and surplusing rate are varied to achieve the consent parameters without excessive power consumption. The process has been studied widely, and mechanisms have been described and quantified for the biochemistry and kinetics of the process. Simple empirical tests have evolved to predict and monitor the performance of the plant which, together with the skilled operator's expertise, can help to maintain good-quality final effluents.

9. TERTIARY TREATMENT

GENERAL CONSIDERATIONS

The secondary treatment processes (using biological filters or activated sludge) should be capable of producing final effluents containing a SS concentration of 40 mg/l and BOD of 30 mg/l. Many conventional works are capable of producing significantly better effluents for long periods as a result of good operational practices, increased resources of power, and monitoring – especially if the incoming sewage has no significant variation in industrial effluent load. However, the Control of Pollution Act 1974 and the Urban Waste Water Treatment Directive[5] have resulted in more stringent standards and compliance criteria. In order to achieve a 95 percentile standard of 30:20, a mean performance of about 15:10 is required. Standards of 15 mg/l SS, 10 mg/l BOD and 5 mg/l ammoniacal nitrogen are often applied and, to achieve a 95 percentile compliance, tertiary treatment is necessary.

The need to remove phosphorus and nitrogen compounds from effluents discharging to sensitive waters has increased the rate of development of new processes and technologies to achieve these stringent effluent-quality requirements.

Conventional sewage treatment has been concerned with solids reduction and BOD removal and, until recently, the reduction of faecal coliforms during treatment was not of great concern. However, water companies are increasingly installing disinfection to sewage discharges into coastal waters[12]. To achieve satisfactory dilution or treatment will usually require disinfection, filtration or irradiation.

The increasing demand in water consumption, coupled with reduced river flows in drier parts of the UK, indicates that a tertiary treatment or 'polishing' stage may be needed at many works. Before adding tertiary treatment to an existing works, a comprehensive survey of the treatment processes, their operation and efficiency, is required so that the optimum design of tertiary treatment plant is achieved. It is generally understood that tertiary treatment processes are most suitable for dealing with well-oxidized effluents and should not be a substitute for inadequate treatment in the secondary stage.

Primary and secondary treatment may remove up to 95% of the BOD, 95% of the SS, 80% of the COD, 90% of ammoniacal nitrogen and 10-40% of total phosphorus, as well as up to 98% of micro-organisms, and can increase the nitrate content from zero to about 30 mg/l.

TERTIARY TREATMENT METHODS

There are many tertiary treatment methods and, whilst some are detailed below, there are many others using similiar or different technologies which are successfully applied. The BOD and SS from a secondary treatment process can be significantly reduced by further biological or filtration systems. The phosphate and nitrate nutrients require more technical biological or physico-chemical methods and significant process changes. A reduction in the micro-organisms requires cell destruction by disinfection or physical separation using very fine filters. Residual or 'hard' COD and colour removal generally require physico-chemical methods.

Lagoons

When an effluent passes through one or more lagoons, the solids settle and a certain amount of biological oxidation occurs. Normally 25-70% of the SS and 15-70% of the BOD can be removed. However, lagoons can suffer from rising and floating sludge, and algal growths may cause a seasonal increase in suspended solids in the final effluent.

Upward-Flow Clarifiers

The clarifier originally consisted of a shallow bed of pea-sized gravel supported on a perforated grating, positioned near the surface of an existing or new humus tank or in a separate unit. Plastic mesh and fine metal (wedgewire) are increasingly common and offer advantages in ease of installation and cleaning. The effluent flows upwards through the bed or mesh where the solids are flocculated and accumulate in the interstices or on the surfaces, possibly dropping down through the tank to the sludge on the tank floor. At intervals the solid matter in the clarifier is removed by backwashing, which involves lowering the water level in the tank so that the whole of the clarifier is exposed, and then washing with a jet of water or effluent. Usually 40-60% of the SS and 20-30% of the BOD is removed.

Rapid-Gravity Sand Filters

The early rapid-gravity sand filters consisted of a layer of graded sand supported on layers of graded gravel resting, in turn, on a special filter floor in which is set a series of nozzles for collecting the filtrate, each unit being held in a suitable casing. It is now common to use several layers of different graded sands and anthracite for the upper part of these filters. The effluent flows down through the filter at a controlled rate. As filtration proceeds, the loss of head increases because of the accumulation of suspended matter. At a certain time each day, or when the head loss has increased to about 3.5 m, the filter is backwashed using filtered effluent. Air-scour is usually combined with backwashing to ensure separation of sludge from the filter particles. The backwashings, containing the sludge, are returned to the inlet of the works.

Moving-Bed Filters
The utility of graded sand to effectively filter secondary settled solids, coupled with a long experience in the water industry, has resulted in a number of alternative filtration methods. Moving-bed sand filters (such as the Dynasand and Toveko systems) remove the sand from the base of the tank using an air-lift pump which both removes, air scours and re-introduces the cleaned sand back into the filtration zone.

Microstrainers
A microstrainer consists of a drum, closed at one end and partially immersed in the flow of secondary effluent, revolving on a horizontal axis and covered usually with a stainless-steel fabric of special weave. Secondary treatment effluent enters the open end of the drum and flows out through the fabric, the solids being retained on the inside of the screen. As the drum rotates, these solids are continually removed from the fabric by strained effluent which is pumped under pressure through a row of jets fitted on top of the machine. Washwater containing the removed solid matter is collected in a trough inside the drum and returned to the works' inlet. Microstrainers are now rarely used at sewage works.

Irrigation over Grassland
Irrigation of the effluent over grass plots is a relatively cheap and effective method of removing solids. The land is divided into two or more plots, one or more of which are in use at a time. The effluent is run onto the plots through a system of channels, flows over the surface and is collected by a second series of channels. The grass needs cutting but should not be kept short. Periodically the plots must be dried off and accumulated solids removed. The suspended solids, and therefore part of the BOD, are retained by the vegetation and the soil. Reductions in SS of 60-80% and BOD of 50-75% can be expected.

Constructed Wetlands
Whilst overland and vertical-flow wetlands achieve up to 95% reduction in BOD and SS, together with significant bacterial reduction, the most popular form of constructed wetland in the UK is the reedbed which consists of gravel-filled tanks planted with

Phragmites australis (the common reed). The tanks are often simple structures comprising fairly shallow excavations lined with an impermeable membrane and having stone gabions at the inlet and outlet ends (Plate 7).

Plate 7. New reedbed at Norton Lindsey, near Stratford-on-Avon

Nutrient Control

The main nutrients responsible for stimulating algal growth are nitrogen and phosphorus. Ammonia is toxic to many aquatic organisms, and concentrations as low as 0.01 mg/l may kill fish. In addition, because the oxygen demand of ammonia is about four and a half times that of carbon, ammonia can produce severe depression of dissolved-oxygen levels. Ammonia and nitrate also affect the re-use of water for potable supply. Fortunately, ammonia and nitrate removal are relatively easily achieved. Because of the chemistry and biochemistry of nutrient removal, there is a need for careful process monitoring and control as well as operational expertise.

Nitrogen Removal. Most conventional works are able to oxidize most of the ammonia to nitrate without difficulty. Where further nitrification is required, treatment in nitrifying filters or biological aerated filters (which are a hybrid between biological filters and the activated-sludge system) may be possible. Reduction of nitrate to allow the escape of nitrogen gas is achieved in anoxic zones where the dissolved oxygen is allowed to reduce sufficiently for the denitrifying organisms to break down the nitrate for their metabolism. The carbon source, which is necessary for these micro-organisms to operate, is provided by the settled sewage, and denitrification is achieved by returning the final effluent into an anoxic zone created in an activated-sludge tank.

Phosphorus Removal. The physico-chemical removal of phosphorus, using iron or aluminium salts to precipitate an insoluble phosphate, is a relatively simple addition at some works. The metal salt may be added at the primary, secondary or tertiary stage but will produce further volumes of sludge. The biological option for phosphorus removal uses the ability of certain bacteria, in the absence of nitrate and oxygen, to metabolize volatile fatty acids and then, upon reaching the aerobic zone of the reactor, are able to absorb all the available orthophosphate. The absorbed phosphate is removed with the surplus activated sludge.

Disinfection

Disinfection is the term describing the elimination or reduction of pathogenic organisms to acceptable concentrations. Complete sterilization is unecessary and would require costly treatment and monitoring. The four chemical techniques using chlorine, peracetic acid, ozone or lime are each capable of giving satisfactory disinfection. However, the choice of method will depend upon many factors such as safety, formation of by-products, capital and running costs, size of plant, etc. Careful consideration of the nature and use of the receiving water, together with Environment Agency consent, are required when evaluating alternative chemical methods.

Physical Separation of Bacteria. Disinfection also refers to the physical separation of bacteria from the secondary or tertiary effluent. Because bacteria are about 1µm in diameter, a very small pore size is needed for effective capture. Whilst reverse osmosis and ultrafiltration will remove inorganic ions and most viruses respectively, micro-filtration is sufficient to achieve the standards required and at a reasonable cost. The cross-flow filtration technique uses a cylindrical porous membrane through which the effluent is pumped, so that micro-filtered effluent permeates the cylinder walls, leaving behind the bacteria and solids which are partly flushed away by the pumped liquid for return to the works' inlet. Backwashing is required to prevent the membrane blinding. The MEMCOR system uses bundles of hollow fibres with the effluent passing from the outside to the inside of the fibres with air used for the backwash cycle. The RENOVEXX system uses a curtain of tubes of high-porosity fabric on which is deposited a filter aid (such as kieselguhr) which forms a layer on the fabric and acts as the filtration membrane. Blinded tubes are cleaned by a squeezing action from the outside, which loosens the solids for return to the works' inlet.

The Kalsep system uses a bundle of fibres which are twisted together, thus offering small pore sizes. When the pores are blinded, the fibres are untwisted to release the filtered solids. The increase in attention to bacterial pollution is leading to a continuous development of micro-filtration methods.

Ultraviolet Irradiation. Ultraviolet irradiation is the method favoured by the Environment Agency, mainly due to the lack of detectable by-products and the extensive work carried out to prove its efficacy. It is most efficient and cost effective when applied to better-quality effluents where turbidity and suspended solids do not interfere with the ultraviolet penetration and absorbance by bacterial cells. The ultraviolet light interferes with the chemical structure of bacteria, disrupting the DNA (deoxyribonucleic acid) and causing death or prevention of reproduction.

Hard COD and Colour Removal

The COD of domestic sewage can be reduced relatively easily by conventional sewage-treatment methods. However, certain industrial effluents will have relatively intractable COD and it may be preferable to remove, isolate, or treat the concentrated waste at the point of production. Coloured final effluents will generally occur when an industrial effluent is discharged from a process where colour-fast or resilient dyes are used. The method of colour removal will depend upon the nature of the residual colour and the results of experimental work with removal methods such as aluminium salts, pH adjustment, granular activated carbon, or ozone. In some cases it may only be possible to reduce the colour intensity.

10. SLUDGE TREATMENT AND DISPOSAL

SLUDGE CHARACTERISTICS

The treatment of sewage produces large quantities of sludge, and the hygienic and safe disposal of sludge is an essential corollary to the treatment of the liquid portion of the sewage. Sludge comprises 'primary' sludge deposited in the sedimentation tanks, 'secondary' sludge resulting from treatment in biological filters or by the activated-sludge process, and 'tertiary' sludges from the polishing processes. Sludge is a thick slurry containing the solids from the sewage dispersed in many times their weight of water. These solids have a considerable affinity for water, and this property is particularly marked with secondary sludges. Under anaerobic conditions untreated 'raw' sludges rapidly degrade, causing bad odours and dangerous gases such as hydrogen sulphide and methane.

For many years, sludge has been disposed of to farm land, sacrificial land, sea or by incineration as a raw, dewatered or digested sludge. There are many processes and methods for the treatment and disposal of sludge, reflecting the particular site and the preferred technology at that time. Because of the difficulties of treatment and disposal, the total cost of sludge treatment and disposal may be up to 50% of the total cost of operating a treatment works. The formation of water authorities (in 1974) introduced a number of regional sludge treatment centres and resulted in an increase in sea disposal for raw and treated sludge. However, the international, environmental and political decisions to ban sea disposal (in December 1998) has significantly influenced the choice of disposal method, with improved disposal to land and incineration being the present preferences. Sludge disposal methods and final disposal points will probably be subjected to regular review; designers and practitioners should therefore be aware of future developments, and if possible, design flexibility into treatment and disposal schemes where this does not significantly increase the cost of projects. The main reasons why sludge has to be treated are to (i) render the sludge inoffensive to achieve minimum impact at the final disposal point, (ii) reduce the volume to be transported off site, and (iii) reduce pathogens. The treatment and disposal of sludge requires extensive processing and handling because it is a concentrated biologically unstable slurry, and it is difficult to separate the solids from the liquid. The mass of dry solids is about 60 g/person/day and produces a raw sludge containing 3-6% dry solids, which is equivalent to about 2.3 l/person/day compared to the 160 l/person/day total water consumption. It is against this background that the many methods of treatment and disposal have evolved.

The solids in the raw sludge contain 70-80% organic and volatile matter derived from fats, grease, food residues, faeces and paper; and the residual inorganic matter mainly consists of siliceous grit. Industrial effluents can significantly affect the nature of the sludge, particularly discharges of heavy metals from metal-finishing processes, plus grit and vegetable matter from food-processing plant. The sludges from secondary and tertiary treatment do not normally settle as readily as primary raw sludge; if they are returned to be settled with the primary sludge they may lead to a lower dry-solids content, and may sometimes cause rising sludge in the primary tanks.

A simple process view of sludge treatment would be to allow the sludge to remain in the primary sedimentation tank for sufficient time to achieve a concentration of more than 10% dry solids, thus reducing the need for process capacity at other stages. However, because of the putrescible nature of sludge and the effect upon the primary tank effluent of an unstable sludge, and the often small diameter of sludge-withdrawal pipes, the raw sludge is withdrawn from primary sedimentation tanks on a frequent basis at about 3-6% dry solids. It is then pumped to a holding tank where separated water may be withdrawn, or ideally to a thickening tank (as described below).

SLUDGE TREATMENT

A much greater amount of the water may be removed by dewatering processes, especially if the sludge is 'conditioned' with chemicals, or other means, to make it separate more readily from its water.

The main methods of sludge treatment are (a) thickening, (b) anaerobic digestion, (c) dewatering, and (d) incineration, and the most common treatment processes throughout the UK are outlined below.

Sludge Thickening

The thickening process may be carried out in holding tanks fitted with draw-off valves at various levels so that, regardless of the depth of sludge in the tank, the top water may be drawn off as the sludge settles and consolidates. The liquors, which may become trapped between settled sludge and floating sludge, may also be withdrawn. The simplest form of thickener is generally operated as a batch process. An improved form consists of a circular tank in which gentle stirring (with a framework known as a picket-fence thickener) encourages the entrained water to leave the slurry and allows the sludge to consolidate in the tank. This thickener can be operated continuously, thus reducing the impact of strong liquors returning to the works' inlet. This process can produce sludges containing up to 9% dry solids. The tendency for some sludges to release gas and cause rising sludge is harnessed in the 'dissolved-air flotation' process, in which a saturated solution of air is introduced into the sludge so that the combined effect of a polyelectrolyte on the sludge and the formation of fine bubbles of rising air, causes the flocculated sludge to rise to the surface from which it is skimmed.

Anaerobic Digestion

Domestic and other organic wastewater sludges are unstable and rapidly become anaerobic as the processes of natural decomposition take place. In the anaerobic sludge digestion process, which is carried out in large closed tanks, the process is enhanced by heating the sludge to about 35°C and excluding air from the system. Efficient mixing enables the daily injection of fresh raw sludge to be rapidly mixed with the digesting mass so that the breakdown of organic matter into soluble and gaseous products is carried out under optimum conditions. As with almost all biological processes, the rate of reaction is approximately doubled every 10°C rise; therefore heated digestion presents process cost savings and throughput increase. The most common range of digestion temperatures in the UK is in the mesophilic range (30-35°C). Whilst enhanced digestion may be achieved in the thermophilic range (40-50°C), at present it has not found favour in the UK.

The complex biological and biochemical processes taking place during heated digestion have been studied widely[13] and involve the conversion of carbohydrates, proteins and fats etc., into simpler soluble compounds including the production of gas having a composition of about 65% methane and 35% carbon dioxide. The gas is collected and burned to produce heat which is transferred into the sludge using one of a variety of internal or external heat exchangers so that a stable temperature, and therefore a stable microbiological system, is maintained. A typical digestion plant (Plate 8) will consist of a large concrete tank partly submerged in the ground with an array of connecting pipework to allow raw sludge to be added, digested sludge to be withdrawn at various levels, sludge gas to be removed, and the digesting mass to be withdrawn and returned after passing through a heat exchanger which utilizes the significant heat energy in the sludge gas (23 000 kJ/m^3)[8]. The digested sludge, which is produced after about 16 days' retention, has a slight tarry odour and is stable. The

Plate 8. Sludge digestion plant showing glass-coated steel tanks with gas mixing and plastic membrane gas-holder (in background)

gas which is evolved may be stored in a 'floating roof' or a separate gas-holder of sufficient capacity to help maintain the temperature during cold winter days. Most conventional digestion systems use a boiler to heat water which then transfers the heat to the sludge, either inside the digestion tank or in an external heat exchanger.

The digestion process is sensitive to a number of variables including heat loss, inhibiting substances, detergents, rags, grit and dry solids of feed. Whilst the anaerobic biological mass offers a diluting and resilient effect on these variables, the skilled operator should regularly monitor the main parameters of temperature, gas yield, volatile acids, sludge feed, total solids and pH, to predict trends in performance.

There are many combined heat and power units (CHP) units installed at both new and old works, because a normal boiler only collects the heat radiated or conducted from the burning gas. About 60% of the energy is lost in the exhaust gases and the unharnessed expansion of the burning gas. The CHP unit extracts the electrical power by burning the gases in an internal combustion engine and provides thermal energy by extracting heat from the engine and exhaust products, which is then used to heat the digestion tank. The electrical power which is generated may be fed into the electrical supply to the works to defray operating costs, and at some works will produce an income if sold to the electricity utility.

The sludge which is produced in the digestion tank needs time to cool down, reduce biological activity and release the dissolved and occluded gases. This is achieved by a short period of storage in a 'secondary' digestion tank.

Whilst the digestion process may produce gas, heat and power, the primary purpose has always been to produce an inoffensive sludge which is acceptable to a variety of final disposal outlets and at a reduced volume. The digestion process has many facets, and it is expected that the evolution of improved systems, equipment and process control, will continue.

Aerobic Digestion

In contrast to anaerobic digestion, aerobic digestion uses aerobic micro-organisms to both stabilize and partly disinfect the sludge. This simple process involves a reaction tank in which air is fully mixed with the sludge, and the subsequent bio-degradation generates a temperature of 50-70°C. At this range of temperature, organic solids are rapidly digested, pathogens are significantly reduced and sufficient heat should be produced to maintain the process. The process has been successfully applied in Europe and the USA, but it has been used at only a few sites in the UK. The plant is inexpensive to build and maintain, and does not require post-digestion storage. However, the energy costs may be high, the off-gas (which is mainly carbon dioxide) may need odour treatment, and there may be difficulties in balancing the sludge availability and the heat balance of the system. Nevertheless, the process is technically and operationally successful – especially for smaller works.

Dewatering

Drying Beds. The earliest form of dewatering used drying beds which were low-walled areas with a top layer of sand, below which layers of gravel, clinker or ash rested on drainage tiles. The water was decanted from the top and the remainder drained through the layers, leaving the sludge to air-dry. Weather problems, the large area of land and labour intensiveness, coupled with improvements in mechanical dewatering, have led to their demise. At some sites they are used for emergencies or dug out and used as sludge storage tanks.

Sludge Conditioning. The sludge solids' strong affinity for water has resulted in the development of chemicals which help to separate sludge from the water. Fortunately, the chemicals are readily available and inexpensive, and include polyelectrolytes which have largely replaced lime and the inorganic salts of iron or aluminium. The conditioning agents assist the solids to coagulate and separate from the liquor layer, thus improving the efficiency and speed of the subsequent dewatering process.

A reduction in the amount of conditioning chemical required is achieved by 'elutriating' or washing the sludge with up to five times its own volume of final effluent and then allowing it to settle. The supernatant liquor removes ammoniacal compounds and reduces the alkalinity of the sludge, as well as some of the fine suspended solids.

Filter Presses. Filter pressing has been used throughout the UK since 1880. The sludge is pressed in chambers between plates lined with synthetic fibre cloth (Plate 9). The water is forced through the cloths and the solid matter is retained in the form of a moist cake which, when the press is 'opened', falls into skips or via a conveyor belt to a storage area. Filter pressing is a batch process relying upon a well-conditioned sludge and pressures up to 15 bar with cycle times of 6-12 hours, depending upon the nature of the sludge and its readiness to release water. The

Plate 9. Filter press

resultant cake, which contains 30-40% dry solids, is the driest form of sludge normally encountered. Because of the basic utility and simplicity, the filter press has remained unchanged except for improved safety, some mechanization and automation, and the introduction of a compressed–air operated membrane which squeezes the sludge cake to produce a drier product and shorter operating cycles.

Centrifuges. Whilst centrifuges were first used on sludge in the mid-nineteenth century, other dewatering methods were preferred until the late 1960s when polyelectrolytes were introduced as conditioning agents and enabled the centrifuge to produce a good solids recovery and a dry cake. Centrifuges are a high-energy, high rotating-speed dewatering method and, because of their small 'footprint', can be easily installed. Centrifuges are capable of producing sludges containing 25-35% dry solids and normally consist of a horizontal cylinder revolving at high speed. Sludge is fed into the cylinder and sludge solids are thrown outwards onto the cylinder surface; a separate revolving screw moves these solids to discharge at one end of the cylinder and the liquid is discharged at the other end.

Belt Presses. This continuous-belt system has operated since 1874 with gradual improvements in operation and performance. Modern belt presses depend upon the particular properties of polyelectrolyte conditioners which encourage formation of large flocs within the sludge, allowing the water to drain away by gravity through porous woven polyester belts. Further dewatering is achieved by compressing the flocs between two porous belts as they pass through a series of decreasing diameter rollers. The belt press does not have the powerful compressing action on the sludge cake which some processes produce, and hence the dry-solids content can vary between 16 and 35%.

Mechanical methods of dewatering are entirely independent of the weather, are flexible in operation and require only a small ground area. However, because of the need to condition the sludge, a significant chemical cost is involved. After dewatering the sludge may be tipped, used on agricultural land, or fed to incinerators. All water which is removed from the sludge is contaminated and is usually returned to the works' inlet for treatment with the incoming sewage.

Incineration

Incineration produces the smallest residual volume of material for final disposal because it evaporates the moisture and oxidizes most of the organic matter into carbon dioxide. The process leaves an ash which is sterile and consists mainly of silicon and aluminium oxides with smaller amounts of various heavy metals, depending upon the industrial effluent discharges. Consequently, care must be taken with landfill disposal. Sludge incinerators need a dry-solids input of at least 30% so that they can operate autothermically to reduce the need for supplementary fuel, and this is assisted by waste-heat recovery from the incinerator exhaust gases.

There are about twelve sludge incinerators in the UK and it is expected that the existing 10% of sewage sludge (which is presently incinerated) may be increased as alternative methods are examined and developed because of the cessation of sea disposal in 1998. Whilst multiple-hearth furnaces were used in the 1970s, fluidized-bed incinerators (Fig. 6) are now established as the main design for sewage sludges.

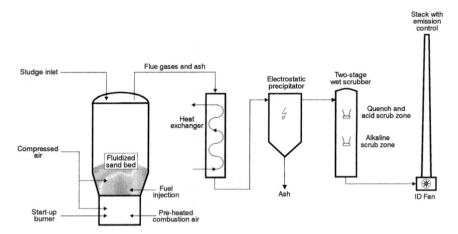

Fig. 6. General arrangement of a fluidized-bed incineration plant

The oxidation of sludge is achieved in a bed of sand which is fluidized by pre-heated air at 600°C. The sludge cake is fed into the incinerator where it mixes with the hot sand, and combustion occurs, which allows the fluidized bed to be controlled at about 850°C. A refractory lining prevents damage to the steel outer shell. The rapidly moving sand particles, coupled with the exchange of heat and intimate mixing with hot air, ensures a rapid combustion of the sludge. The fluidized bed has no moving mechanical parts and, because the combustion efficiency is high and the temperatures are uniform, the residence times for complete combustion are reliably achieved. The rapidly moving sand also provides continuous attrition of the burning material, removing the layer of char as it forms and exposing fresh material for combustion[14]. The exhaust gases contain ash which is incombustible, nitrogen oxides because of the 3-5% of nitrogenous content in the sludge feed, sulphur dioxide which may be produced from organic matter and some industrial effluents, and water vapour. The ash and particulate matter may be removed by electrostatic precipitators, wet scrubbers or bag filters. The acidic gases, including sulphur oxides and some nitrogen oxides, can be absorbed in wet scrubbers containing caustic soda. The nitrogen oxides may need to be further controlled by catalysis or decomposition using direct injection of urea or ammonia. The performance of incinerators is strictly monitored and controlled so that complete combustion occurs and to ensure that the gaseous emissions comply with current regulations.

SLUDGE DISPOSAL

Disposal at Sea
Until the cessation of sea disposal in December 1998, conveyence in specially constructed ships was an acceptable method of disposal. This was practicable only on a big scale for towns near the coast, e.g. London, Edinburgh, Glasgow, Newcastle and Manchester, and was controlled by the Ministry of Agriculture, Fisheries and Food.

Lagooning
A temporary method of dealing with liquid sludge is to pump into lagoons formed by excavation and banking. Alternatively, the lagoon may be lined with pre-cast concrete slabs and provided with valves and floating arms to withdraw any water which separates. Lagoons are relatively cheap, but the sludge may still have to be disposed of at a later date, and odour problems and health and safety risks can arise.

Disposal to Land
For many years raw and digested sludges were transported and sprayed on farmland with limited control of hydraulic overload, runoff to watercourse, toxic metal content, smell nuisance (particularly with raw sludge), and contamination of crops or livestock. A significant improvement was achieved by the introduction of subsoil injection which greatly reduces the visual and smell nuisance and allows the land to be returned to use earlier. Raw sludge may be injected into grassland or may be sprayed onto arable land provided that it is ploughed in. Sewage sludge may also be used effectively in forestry and woodland. The reclamation of derelict land, such as that caused by coal mining and mineral extraction, can be achieved by application of sludge.

There are many restrictions on the application of sludge to land involving European and UK regulations, the Environment Agency, and codes of practice which limit the way in which the sludge may be applied to various types of land or crop types.

Domestic sewage sludge is an exploitable resource and should be recycled for beneficial use wherever possible, subject to the appropriate safeguards, because this is the preferred environmental option (Chapter 11).

If it is not practicable to utilize sludge due to the concentration of metals or other contaminants or where the total cost of recycling would be uneconomic, or there is a need for a short-term expedient, the disposal of sludge to landfill sites may be appropriate.

The expected continual increase in Landfill Tax, and the reduced availablity of suitable sites, will increase the cost of disposal to landfill sites.

11. UTILIZATION OF PRODUCTS OF SEWAGE TREATMENT

The gradual acceptance that many of the earth's resources are finite, coupled with the increased pressure on resources, has focused the approach to recycling and reduction of waste. During the last decade, the apparent increase in drought conditions in parts of the UK has raised the environmental and political interest in water leakage, diminished river flows, and damage to the natural environment. It is now common to recycle glass, paper, aluminium and some plastics. The move towards designing recyclable components of automobiles and other large-volume consumer items is now the preferred way to manage resources. Because the water industry produces a liquid and solid waste which had to be reasonably acceptable for return to the environment, the water industry was a forerunner in the conservation, re-use and utilization of by-products.

The gradual move towards higher standards of effluent discharged to rivers, thereby increasing the environmental benefits, also produces social and recreational benefits.

Whilst sewage is generally returned to the environment at a considerable distance from the initial abstraction point, with a consequent reduction in both surface and groundwater between source and point of use, the discharge of a good-quality effluent can return much of a river system to its former uses. The downward percolation of the increased river flow will also help to recharge aquifers. The treated effluent and river-water admixture is also available for navigation, and at suitable points downstream may once again be abstracted for industrial or agricultural purposes or for treatment and re-use as a public water supply. The discharge of high-quality effluents will also increase the potential of a river to augment another river by direct transfer.

HEAT AND ENERGY FROM SLUDGE GAS

Anaerobic sludge digestion is a method of treating sewage sludge to reduce the final volume and make it acceptable for disposal. The methane gas mixture which is produced should have sufficient calorific value to maintain the digestion process at 35°C throughout most of the year. Where works were well designed and operated and located in warmer climates, surplus gas has been used to heat buildings and greenhouses. Some large, old works used dual-fuel engines to produce electrical power as well as using the heat generated to maintain the digester temperature. These engines were robust but required a fuel-oil injection of about 4%. They are becoming obsolete with the increasing introduction of spark-ignited combined heat and power (CHP) engines which have greater efficiency and considerably less manning requirements. CHP units using sludge gas may be used in modules as small as 20 kW power output and up to 1.0 MW (Plate 10). There are more than 150 sites in the UK which have CHP units capable of producing a total power output of greater than 95 MW. The short pay-back periods of three to four years or less make CHP an attractive option at both old and new digestion plants[15]. Maximum benefits for CHP are achieved when particular attention is paid to good design, dry-solids loading, plant insulation and heat-recovery systems. Whilst problems may be experienced with hydrogen sulphide, which is oxidized to sulphuric acid and causes increased corrosion of engines and exhaust systems, the benefits of overall greater efficiency and significant savings in site electrical energy costs, coupled with significant sales of electrical power to the electricity utility, make this an attractive process.

Plate 10. Combined heat and power installation

SOIL CONDITIONING

Sewage sludge offers many benefits to the farmer if the limitations are known, together with an understanding of the needs and nature of the soil and the type of sewage sludge. Although sludge does not offer the same balance, consistency and availability of nutrients as a manufactured inorganic fertilizer, it has been applied to land on a large scale for many years. It is beneficial to crop production and can provide a significant saving in fertilizer costs. Sludges can be sprayed, spread, and injected. The concerns regarding risks to health from pathogens, and the progressive contamination of land with heavy metals, have been addressed by regulations and codes of practice which reflect the national importance of sludge utilization in agriculture. Domestic sewage sludge, when delivered to site by a water company, provides a well-controlled and documented soil additive which will increase the amount of slowly available nitrogen and phosphorus as well as fibrous and organic material to 'condition' the soil, thus creating a more open texture which increases its water-holding capacity. Potassium is not as abundant as phosphorus and nitrogen, and may have to be supplemented where sludge is used as the main soil additive. Most of the micro-nutrients which are essential for plants and animals (including boron, selenium and molybdenum) are usually present in sewage sludges. There is a financial benefit to the farmer as well as a reduction in disposal costs. The final disposal point for sludge in both the short and long term should be a major consideration when deciding the treatment processes to which the sludge will be subjected. The introduction of the Landfill Tax (to reduce the amount of material unnecessarily disposed of to landfill) has increased the drive towards beneficial use for sewage sludge. Composting sludge with straw, woodchips or domestic refuse has been tried many times in the past and may gain favour as an alternative in the future. The potential problems of pathogens, heavy metals, leachate, product handling and soil conditioning are addressed by processes which subject the sewage sludge to a finally divided form of calcium oxide which, as it hydrates, generates temperatures of above $50°C$ and a pH of about 12, thus significantly reducing the pathogen content and the solubility of heavy metals.

HEAT AND ENERGY FROM SLUDGE

Many water companies should be able to achieve the economic and process benefits of increased regional sludge treatment following the cessation of sea disposal in December 1998. The economy of scale produces benefits in the unit costs of treatment and disposal, allows novel processes to be considered and the benefits of the inherent qualities of the sludge to be exploited. In some parts of the UK, the use of sludge as a soil conditioner or in final disposal as landfill may not be viable options, therefore sludge drying to produce a pelleted product may be an attractive option. Any pelleted dried sludge which cannot be used may be landfilled without many of the difficulties associated with sludge cake and, as a dried granular material has a significant volumetric reduction, there are cost savings.

A dried product may be used in many ways. In agriculture it can be used as a soil conditioner and as a slow-release nitrogen fertilizer. The pellets are easily stored and can be spread using conventional farm machinery. If a temperature of 400°C is maintained in the drying process, the pathogens should be reduced to levels naturally occurring in soils.

Dried sludge can also be used as a fossil fuel substitute because of the high concentration of organic carbon compounds. In a dried sludge (especially an undigested sludge), the calorific value of the material is readily available in a usable form and the energy can be used in many ways; for example (a) power generation, (b) brick, cement or steel manufacture, (d) district heating, (e) low-temperature pyrolysis to produce a fuel oil, or (f) high-temperature gasification to produce a fuel gas.

The gasification of the dried undigested sludge at a temperature of about 800°C is a further development in the search for improved sludge utilization processes. The basic technologies used are well-established processes in other industries. Gasification produces a medium calorific-value fuel gas which, after cleaning and compression, can be used to fire a CHP gas turbine to generate electricity. If the heat from the CHP plant can be used in the sludge-drying process, the system can be very energy efficient. There is also a 70-80% volumetric reduction of the dried sludge in the gasification process, leaving an ash residue which is easily landfilled.

Demonstration plants gasifying sewage sludge will probably be seen within the next five years. These plants will prove the technology and will determine the operating parameters and process adjustments which are needed to maximize the economic and environmental benefits of the process.

It is expected that the innovative approach to sludge treatment, utilization and disposal will continue as compliance with the Urban Waste Water Treatment Directive[5] effectively increases the amount of sewage sludge which is produced.

12. CONTROL, TRAINING AND ADMINISTRATION IN UK

CONTROL OF SEWAGE-TREATMENT PROCESSES

The improvement in treatment processes and the drive for higher-quality effluent discharges, coupled with the need for greater reliability of processes and rapid response to plant changes, has stimulated the adoption of modern controls, automation, mechanization, instrumentation and telemetry. Whilst newly built treatment works will incorporate many modern control and measurement systems to achieve the work's performance required, there are many sewage-treatment works throughout the UK which rely upon relatively simple processes to achieve treatment. New works may have sophisticated controls for dissolved oxygen, flow and sludge withdrawal; however, the performance of many older works may be significantly improved by the addition of simple sensors and alarms to detect malfunctioning and performance changes with telemetry to a central control where remedial action may be taken or the fault condition relayed to the operators by telemetery. At large works, distributed or centralized computerized data-handling systems allow the operation of certain parts of works to be automatically controlled within pre-set parameters together with a historical log for later interrogation.

Highly sophisticated laboratory analytical techniques and instruments are used to examine sewage, industrial effluents and treated effluents together with considerable scientific effort to maintain effective treatment processes involving chemists and biologists working closely with engineers and managers. The analytical results are used to make changes to process operations to achieve the required performance and cost-effective use of resources. The control of industrial effluents to avoid toxicity or excessive loads discharged to the sewage-treatment works, and the need to reimburse the water company for the conveyence, treatment and disposal of industrial waste, relies upon accurate analysis of the contributing industrial and domestic waste.

The charging formulae and unit costs will vary throughout the UK, depending upon the characteristics of process plant used, domestic sewage treated, and the costs of treatment[7].

Staff Training and Assessment

Because of the greater emphasis on process control to achieve the stringent final effluent standards, the complexity of some systems and the need to respond quickly and effectively to changing conditions, the staff operating sewage-treatment works need adequate training. CABWI is an awarding body for water and sewage National Vocational Qualifications (NVQs)[16]. The corresponding Scottish Vocational Qualifications (SVQs) are approved by the Scottish Qualifications Authority (SQA)[17]. Many of the water companies, Scottish water authorities and DoE (Northern Ireland) Water Service have taken advantage of these qualifications to upskill and develop their employees. To achieve an S/NVQ award, operational staff must meet national criteria as detailed in each S/NVQ. As well as meeting knowledge and understanding requirements, the criteria require staff to demonstrate their competence by presenting examples of their work to an accredited assessor.

ADMINISTRATION IN ENGLAND AND WALES

The reorganization of the water industry in England and Wales in 1989, provided for by the Water Act 1989, established ten water and sewerage companies – replacing the ten regional water and sewerage authorities. The responsibilities of these companies cover water supply, sewerage and sewage treatment and recreation associated with their land and waters. Some district councils continue to act as agents of water and sewerage companies for certain sewers but not for industrial effluent control or sewage treatment, and the 'water only' companies supply water in their areas. Responsibility for licensing water resources, river pollution prevention and fisheries rests with the Environment Agency.

Dwr Cymru (Welsh Water) includes those parts of the Rivers Wye and Dee in England but excludes those parts of the River Severn in Wales. In England, Thames Water covers the river catchment including the London area, Anglian Water covers eastern England from the Thames estuary to the Humber, Yorkshire Water covers the traditional county area, Northumbrian Water covers North East England to the Scottish border, North West Water from the Scottish border to Cheshire, Severn Trent Water the mainly inland areas of the rivers Severn and Trent, Wessex Water extends from the Bristol Channel to the East Devon and Dorset Coast, South West Water extends from the river Tamar westwards, and Southern Water covers the south east from the Isle of Wight to Kent[18].

ADMINISTRATION IN SCOTLAND

The Local Government etc.(Scotland) Act 1994 provided for the creation of three new public bodies, to be known as the East of Scotland Water Authority, the West of Scotland Water Authority and the North of Scotland Water Authority. From 1 April 1996, the three new authorities took over the provision of water and sewerage services from the regional and island councils which had carried out this statutory function by virtue of duties placed on them by the Water (Scotland) Act 1980 and the Sewerage (Scotland) Act 1968. As public bodies, the authorities are accountable to Parliament through the Secretary of State.

The Local Government etc. (Scotland) Act 1994 also established the Scottish Water and Sewerage Customers Council to represent the interests of customers and potential customers of the new water and sewerage authorities.The Council has a particular role setting charges and improving codes of practice for customer relations, which authorities are statutorily required to produce.

Since 1 April 1996, the river surveillance function throughout Scotland (including the islands) has been the responsibilty of the Scottish Environment Protection Agency which is now the main body for the regulation of all environmental issues in Scotland and also has responsibility for the control of most emissions to air and land[19].

ADMINISTRATION IN NORTHERN IRELAND

The duty of promoting the conservation and cleanliness of the water resources of Northern Ireland is placed with the Environment and Heritage Service – a 'Next Steps Agency' within the Department of the Environment for Northern Ireland, under the Water Act (Northern Ireland) 1972. The 1972 Act also makes provision for the prevention of pollution of waterways, and the Environment and Heritage Service operates a 24-hour pollution response system to deal with reports about water pollution.The Service also monitors river water quality under an extensive programme and is responsible for monitoring bathing water quality in Northern Ireland.

Under the terms of the Water Act, consent to the discharge of industrial effluent or domestic sewage to a watercourse or to underground strata must be obtained from the Environment and Heritage Service, Water Quality Unit in Belfast[20]. At present there is no licensing system governing water abstraction in Northern Ireland.

The responsibilty for public water supplies and sewerage is vested in the Department of the Environment for Northern Ireland under the Water and Sewerage Services (Northern Ireland) Order 1973.

The duties, requirements and powers derived from this legislation are undertaken by Water Service – the Next Steps Agency. In addition to its Belfast Headquarters, Water Service has eastern, northern, southern and western divisions based, respectively at Belfast, Ballymena, Craigavon and Londonderry[20].

REFERENCES

1. ROYAL COMMISSION ON SEWAGE DISPOSAL. *Final Report.* HMSO. 1915.
2. REPORT OF THE WORKING PARTY ON SEWAGE DISPOSAL. *Taken For Granted.* HMSO. 1970
3. CHARTERED INSTITUTION OF WATER AND ENVIRONMENTAL MANAGEMENT. *An Introduction to Wastes Management.* 2nd Edition. CIWEM, 1995.
4. COUNCIL OF EUROPEAN COMMUNITIES. Directive on pollution caused by certain dangerous substances discharged into the aquatic environment of the Community (76/464 EEC). *Official Journal L129/23,* 18 May 1976.
5. COUNCIL OF EUROPEAN COMMUNITIES. Directive concerning urban waste water treatment. (91/271/EEC). *Official Journal L135/40,* 30 May 1991.
6. The Urban Waste Water Treatment (England and Wales) Regulations 1994. Statutory Instrument 2841. 1994.
7. CHARTERED INSTITUTION OF WATER AND ENVIRONMENTAL MANAGEMENT. *An Introduction to Industrial Wastewater Treatment and Disposal.* CIWEM, 1997.
8. CHARTERED INSTITUTION OF WATER AND ENVIRONMENTAL MANAGEMENT. Handbooks of UK Wastewater Practice. *Sewage Sludge: Introducing Treatment and Management.* 1995.
9. INSTITUTE OF WATER POLLUTION CONTROL. Manuals of British Practice in Water Pollution Control. *Unit Processes. Biological Filtration.* IWPC, 1988.
10. INSTITUTION OF WATER AND ENVIRONMENTAL MANAGEMENT. *An Introduction to Water Supply in the UK.* IWEM, 1994.
11. CHARTERED INSTITUTION OF WATER AND ENVIRONMENTAL MANAGEMENT. Handbooks of UK Wastewater Practice. *Activated-Sludge Treatment.* CIWEM. 1997.
12. COUNCIL OF EUROPEAN COMMUNITIES. Directive concerning the quality of bathing water (76/160/ EEC). *Official Journal L31/1,* 5 February 1976.
13. CHARTERED INSTITUTION OF WATER AND ENVIRONMENTAL MANAGEMENT Handbooks of UK Wastewater Practice. *Sewage Sludge: Stabilization and Disinfection.*1996.
14. ENVIRONMENT AGENCY. IPC Guidance note S25.01. Processes subject to Integrated Pollution Control. Waste Incineration, 1996.
15. DEPARTMENT OF THE ENVIRONMENT *Introduction to Small- Scale Combined Heat and Power.* Good Practice Guide No.3 ETSU 1995.
16 CABWI AWARDING BODY. 1 Queen Anne's Gate, London SW1H 9BT.
17. SCOTTISH QUALIFICATIONS AUTHORITY. Hanover House, 24 Douglas Street,

Glasgow G2 7NQ.
18. DEPARTMENT OF THE ENVIRONMENT, TRANSPORT AND THE REGIONS. Romney House, 43 Marsham Street, London SW1P 3PY.
19. THE SCOTTISH OFFICE, AGRICULTURE, ENVIRONMENT AND FISHERIES DEPARTMENT. Environmental Affairs Group, Victoria Quay, Edinburgh EH6 6QQ.
20. DEPARTMENT OF THE ENVIRONMENT FOR NORTHERN IRELAND, ENVIRONMENT AND HERITAGE SERVICE. Calvert House, 23 Castle Place, Belfast BT1 1FY.

INDEX
Italic page numbers denote illustrations

Activated sludge	31–36	Administration	
aeration systems		England and Wales	6–7, 58
diffused air	32	Northern Ireland	59
surface aeration	32–33	Scotland	6–7, 59
comparison with biological		Ammonia	13, 17, 40
filtration	36	Analysis	10, 12, 13, 57
deep shaft	34	Aquatic life	12,13
extended aeration	35		
oxidation ditch	36	Bacteria	17, 25
oxygen activated sludge	34	in activated sludge	31
Unox process	35	in biological filters	26
Vitox process	35	removal	37, 41, 55
Acts of Parliament		Belt presses	49
Rivers Pollution Prevention Act		Biochemical oxygen demand	
1876	5	(BOD)	10, 12
Public Health (Drainage of Trade		removal	37
Premises) Act 1937	6, 8	Biological filtration	6, 25–29, 26, 28
Rivers (Prevention of Pollution)		alternating double filtration	28–29
Acts 1951 and 1961	6	combined processes	29
Rivers (Prevention of Pollution)		comparison with activated sludge	36
(Scotland) Acts 1951 and 1965	6	distributors	26–27
Clean Rivers (Estuaries and Tidal		filter construction	25–26
Waters) Act 1960	6	filter flies	27, 28
Public Health Act 1961	6	organisms	26
Water Resources Act 1963	6	recirculation	27, 28
Sewerage (Scotland) Act 1968	59		
Water Act (Northern Ireland) 1972	59	Captive-film systems	25, 29–30
Water Act 1973	6	Centrifuges	49
Water and Sewerage Services		Chemical oxygen demand	
(Northern Ireland) Order 1973	59	(COD)	12, 13, 42
Control of Pollution Act 1974	7, 37	Cholera	4
Water (Scotland) Act 1980	59	Colloids	15
Water Act 1989	58	Colour removal	42
Urban Waste Water Treatment		Combined heat and power	
(England and Wales) Regulations		plants	46, 54, 56
1994	8	Comminutors	19
Local Government etc. (Scotland)		Composting	55
Act 1994	59	Conditioning of sludge	47–48

INDEX

Conservation	8	Kalsep system	42
Cooling water	9		
		Lagoons	
Deep-shaft activated-sludge		sludge	51
process	34	tertiary treatment	38
Digestion – *see under* 'sludge'		Land disposal	51
Disinfection	41	Land treatment	6, 25
Disintegrators	19	Landfill	52, 55
Dissolved-air flotation	45	Legislation – *see under* 'Acts of	
Domestic sewage	14	Parliament'	
Dry-weather flow (DWF)	21		
Drying beds	47	Membrane filtration	41-42
Dynasand system	39	MEMCOR system	41
		Mesophilic digestion	45
Effluent standards	6, 8, 13, 37	Methane	43, 45, 54
Environment Agency		Microfiltration	41
	7, 12, 15, 21, 41, 42	Micro-nutrients	55
European Union Directives	7		
Eutrophication	17	National Rivers Authority	7
Extended aeration	35	Nitrates	13, 33, 40
		removal	34
Filter presses	*48*	Nitrification	17
Fish kills	5	Nitrogen	13
Fluidized beds	30, *50*	in the activated-sludge	
		process	33–34
Gasification	56	removal	40
Grit separation	20	Northern Ireland	
		administration	59–60
Hard COD	13	Nutrients	13
removal	42	control	40
Heavy metals	55		
HM Industrial Pollution Inspectorate	7	Office of Water Services (OFWAT)	7
HM Inspectorate of Pollution	7	Oxidation ditch	36
Humus tanks	27	Oxygen activated sludge	34
		Oxygen solubility	35
Incineration –*see under* 'sludge'			
Industrial effluents	1, 5, 8, 9, 14, 15, 44	Permanganate value	12
charging	9, 13, 15, 58	Phosphorus	13
Industrial Revolution	5	removal	41

64

INDEX

Phragmites australis	40	Sewage	1, 14–15
Picket-fence thickener	45	farms	25
Pollution	1, 4, 5	treatment works	2, 10, *11*, 15, *16*
Polyelectrolytes	45, 48, 59	control	57
Preliminary treatment	18–21	Sewerage systems	1, *2*, 4, 14
Primary treatment	22–24	Sludge	43–52, *54*–56
Privatization	7, 58	characteristics	43–44
		dewatering	47, *48*, 49
Receiving waters	7, 8, 10, 12	digestion	45, *46*, 47
Reedbed	39, *40*	aerobic	47
Recirculation	27	anaerobic	45–*46*
RENOVEXX system	41	gas	45, *54*
Re-use	8–9, 53	disposal	51–52
Reverse osmosis	41	lagoons	51
River Authorities	6	land	43, 51–52
River Boards	6	sea	51
River pollution	1, 4, 5	heat and energy	54
River Purification Boards	6	humus	27
Rotating biological contactors	*29*, 30	incineration	49, *50*, 51
Royal Commission on River Pollution		primary	22, *23*, 24
Prevention, 1868	5	soil conditioning	55
Royal Commission on Sewage		solids	44
Disposal, 1898	6	surplus activated	31
effluent standard	13	thickening	44–45
		Staff training	58
Scotland		Storm-sewage treatment	20–21
administration	59	Suspended solids	10, 12, 13
HM Industrial Pollution			
Inspectorate	7	Tertiary treatment	37–42
River Purification Boards	6, 7	irrigation over grassland	39
Scottish Environmental Protection		lagoons	38
Agency (SEPA)	7, 15, 59	microstrainers	39
Scottish Water and Sewerage		moving-bed filters	39
Customers Council	59	nutrient control	40–41
Vocational qualifications	58	rapid-gravity filters	38–39
Water Authorities	59	reedbed	39–*40*
Screening	18, *19*	upflow clarifiers	38
Secondary treatment	25–36	Thermophilic digestion	45
Sedimentation	22–24	Toveko system	39

INDEX

Toxic metals	14	Wastewaters	1, 10
		analysis	10, 12, 13, 57
Ultrafiltration	41	Water-carriage system	1
Ultraviolet irradiation	42	Water closets	4
Unox process	35	Water Authorities (Scotland)	59
Urban Waste Water Treatment		Water Quality Unit (N. Ireland)	59
Directive	8, 37	Water Service (N. Ireland)	60
		Water Service Companies	
Vitox process	35	(England and Wales)	7, 15, 58
Vocational qualifications	58		